A REALIST PHILOSOPHY OF SCIENCE

This book is about the nature of scientific theory. The central topic of inquiry concerns how it is that theories are able to supply us with powerful and elegant explanations of puzzling phenomena that often confront the scientist and layman alike. It is argued that an answer to this question supplies us with an account of how theories achieve a variety of tasks such as the prediction and organisation of data, including how they support a very important class of claims known in the literature as counterfactual conditionals. The book begins by presenting a critical survey of past, classic formulations of the nature of scientific theory which are prominent in philosophy of sciences circles today. These include the doctrines of logical positivism, Hempel's Deductive–Nomological model of explanation, Hanson's gestalt approach to understanding and observation, Kuhn's sociology of science, and others.

After presenting the reader with a critical examination of the above approaches to the nature of scientific theory, the author then presents his own views. His approach is essentially an ontological one. Ontology is usually characterised as the study of the nature of the most fundamental constituents of the universe. The major contention of the book is that theories are essentially depictions of the nature of things, and that it is this feature which accounts for their ability to explain, predict and organise a vast array of data. In the tradition of more recent versions of scientific realism that have occurred in the literature, the author attempts to show that the very confirmation of a theory depends on its ability to refer to the fundamental constituents of nature. It is argued that science can function only from an ontological point of view. In order to show this, the student is presented with a model of how theories are confirmed which is then conjoined with a model of the nature of scientific explanation. In so doing, the author ends up fostering a view of science which is rather controversial to twentieth-century philosophical tradition, namely that science is really metaphysics in disguise but a metaphysics which can ultimately be judged by empirical standards. Such an approach to science characterises the modern-day scientist as an old-fashioned natural philosopher.

Jerrold L. Aronson is Associate Professor of Philosophy at the State University of New York at Binghamton. He received his BS in physics at the University of Illinois and his PhD in philosophy at the University of Wisconsin. From 1967 to 1969 he was an Assistant Professor of Philosophy at Tufts University.

He has contributed to scholarly journals on such topics as causation and scientific explanation, the nature of theory reduction and issues pertaining to the philosophy of physics.

A REALIST PHILOSOPHY OF SCIENCE

Jerrold L. Aronson

St. Martin's Press   New York

ISBN 0–312–66474–5

Library of Congress Cataloging in Publication Data

Aronson, Jerrold L., 1940–
   A realist philosophy of science.

   1. Theory (Philosophy)   2. Science–Philosophy.
I. Title.
B842.A76 1983   121   82–22956
ISBN 0–312–66474–5

To my family and friends, who are wonderfully more
complicated than the theories mentioned in this book

Contents

Preface

The major purpose of this book is to present the reader with an account of how theories are able to perform a variety of important tasks such as explanation, prediction, the organisation of data and the support of various other scientific claims. The central intellectual quest, then, is to develop a theory about the nature of theories, especially that aspect which leads to powerful explanation and deep understanding of a variety of phenomena. The book begins by critically examining past formulations of theories that have played a dominant role in the traditional philosophies of science. These include logical positivism, Hempel's Deductive–Nomological model, Hanson's gestalt approach, Kuhn's sociology of science, Salmon's statistical relevance model, and others.

However, as much as I respect these past approaches to the nature of scientific theory, believing that each view has, in its own way, brought out something very important about theories, I nevertheless feel that it is time to promote an entirely different approach. The view of theories I explicate here attempts to do justice to the highly complex aspects of scientific explanation, prediction and observation. In doing so, it is also intended to clarify the value of contributions of the above-mentioned philosophers. After critically examining these approaches, I will go on in the second part of the book to present arguments for my views on the matter with respect to the traditional issues in the philosophy of science, for example, the nature and logic of: confirmation, explanation and prediction, the reduction of theories and counterfactual conditional claims.

The material covered in the text is directed primarily towards the advanced undergraduate or beginning graduate science major who may have had little or no background in philosophy but is nevertheless interested in foundational or methodological problems which occur within a scientific discipline. Likewise, it is my hope that philosophy majors of a scientific bent should be able to relate the discussions in these chapters to traditional philosophical issues. I am assuming that most undergraduate philosophy of science courses will cover the work of highly influential philosophers such as Hanson, Hempel and Kuhn. As the reader can see from the table of contents, Part I of the book is designed as a companion reading to what I consider to be classics in the philosophy of science.

The approach to the nature of theory which I am advocating in Part II of the text is an ontological one. Ontology is usually characterised as the study of the nature of the most fundamental entities of which the universe is comprised, whether they be atoms, fields, quarks, twisters, gravitinos. . . . It is my contention that theories are first and foremost depictions of the nature of things, and it is this feature which accounts for their ability to explain, predict and organise a vast array of data. A corollary of this, of course, is that to the extent a theory has little to say about the nature of its subject matter, among other things, it will lack richness in explanatory power and its predictions will be relatively scarce. It will even be less open to confirmation in comparison to rival theories which speculate more on the nature of things. On my view, then, theories are metaphysical systems which are subject to empirical considerations. To many philosophers and scientists alike, such an opinion is extremely heretical. It can be said that Part II is devoted to justifying the view that science is really metaphysics in disguise, that science can not function except from an ontological point of view.

I am indebted to many individuals who were very instrumental in the development and refinement of the ideas that have gone into this book. Special credit goes to Fred I. Dretske and Haskell Fain who, I believe, set me on the right course during my graduate studies at the University of Wisconsin. I should also like to thank the SUNY Foundation for providing me with summer support and the administration at the State University of New York at Binghamton for travel support which enabled me to test my ideas at other universities. Many, many thanks go to my friends, David A. Edwards, Rom Harré, Larry Roberts, Eileen C. Way and Robert Weingard, for their frequent discussions of the material that went into the text, for their constructive criticism of the original manuscript and for their inestimable intellectual support. Without their encouragement, I doubt that this project would ever have been completed.

J.L.A.

Part I

1 Introduction

How often in the course of our daily thinking are we actively constructing theories about events, people, politics, even about fate or a deity, in order to make sense of what has happened or is presently occurring to us and others? How often do we use these theories as a basis for inferring things about future events? For many of us, there is not one important decision in our lives, whether it be an intellectual or a practical one, which does not in some way rest heavily upon several theories about nature, the psychological–sociological make-up of persons and groups, etc. Except for the 'lucky' ones, those decisions which are based upon unreasonable or false theories very often lead to disaster and unhappiness. It can hardly be unreasonable to believe, although I will not defend such a belief here explicitly, that decisions based upon theories which reflect reality will be more likely to put us on the right path in life.

In light of their prominent role in our lives, it should be of intrinsic intellectual interest to inquire into the nature of theories, to seek an understanding of how they work to bring about explanation and prediction of events, how they can serve to organise our experiences, and so on. The development of a theory about theories, then, will be the major objective of the book. Another related topic is the question of how we can tell if a theory is a good one or not. After all, not only does our practical life demand that we construct and utilise a variety of theories, they should be *good* – whatever *that* means – ones as well. I hope that we will come up with a set of reasonable criteria which enables us to glean the good from the bad. I leave the reader with this warning, however. Coming up with answers to the above questions does not guarantee in the least that we will all become skilled theoreticians, that we will be in a better position to come up with new and powerful theories. Such an expectation is far too pretentious. That would be like believing that becoming an expert commentator on chess strategy automatically makes one a chess player of world championship calibre. Rather, what I go on to say about the nature of theory and the marks of good theory will, one hopes, help students come to a greater appreciation of the power and complexities of the theories they come across in the course of their scientific studies, to experience them for what they really are instead of what certain philosophers and scientists insist they be.

1. PAST APPROACHES TO THE NATURE OF SCIENTIFIC THEORY

Since it is impossible to cover comprehensively all the major approaches to the nature of scientific theory, I will focus my investigations on what I consider to be some of the more classical and related views on the matter, beginning with the school of philosophy known as logical positivism. The positivistic view of science dominated the philosophy of science from the turn of this century, and although it has since been repudiated by almost every school of philosophy, many philosophers can be found who still subscribe to many of its tenets. Since positivism, the philosophy of science has evolved into the formalistic camp which placed almost the entire emphasis on analysing theories in terms of symbolic logic. Like the logical positivists, they felt that the powerful, precise and rigorous system of logic that was developed by Russell and Whitehead in *Principia Mathematica* should be the major tool of analysis of the language of science. This school of thought culminated in Hempel's Deductive–Nomological model. Since Hempel's work, two radically different approaches to theory have emerged. On the one hand, we find a continuation of the formalistic line by those philosophers of science who have attempted to surmount various technical problems that have plagued Hempel's analysis and who have sought to develop the paragon formal model of a theory. But there is another group of philosophers who believe that formal approaches to theory are bankrupt, that we cannot get at the essence of a theory by focusing on its logical structure; instead there are other aspects of a theory that determine its essential nature in such a way that the meaning of a theory varies from non-logical context to context. While these philosophers may differ on what controls such a variation in a theory's meaning, they all agree that the key to understanding the meaning of a theory does not lie in its logical syntax, and whatever meaning it has must be qualified by some historical, semantical, social-political or psychological perspective. For this reason I will label these antiformalist approaches to theory *contextualist*.

Saving the details of these views for the remaining chapters, let me just briefly mention what they are about, making a few superficial comparisons here and there. The logical positivists (sometimes called logical empiricists) felt that any theory could be reduced to nothing but an organised description of observable phenomena or data. Their goal was to make science a purely empirical enterprise, one where the scope of scientific knowledge does not go beyond immediate experience. One result of this extremely narrow conception of scientific theory is that unobservable entities were relegated to being 'convenient myths' whose sole purpose was to aid the scientist in organising data – and nothing more! As years went by, philosophers of science were forced to give up such a stringent view on the existence of entities which could not be directly observed but they still shared with the positivists the belief that scientific theories were first and foremost logico-epistemological systems that were to be understood only by translating scientific language into logical (canonical) notation. The meaning of a theory was essentially a function of its logical syntax along with the class of things or objects to which the (non-logical) terms in the theory refer. Hempel's Deductive–Nomological model was an outgrowth of this particular view.

The contextualists, on the other hand, felt that there was much more to theory than its being a logical calculus which merely enabled the scientist to make predictions. Some philosophers believed that the actual meaning of a theory was to be found in the picture of reality it carried, i.e., theories had formal features but their identity was a function of the *model* of nature contained within. Naturally, the formalists felt that models, being non-logical, 'vague' picturing devices, were or ought to be completely dispensable. If anything, they felt, models interfered with the very elegance of a theory! In fact, the formalist would argue that any non-logical contextual element of a theory is irrelevant to its essence. There are other contextualists who focused, instead, on the psychological-semantical features of a theory; in this case, the essence of a theory lies not in its logical syntax but in the fact that a theory is a *multi-levelled language system* by means of which the scientist is able to *perceive* or *pattern data as a unified whole*. Getting at the meaning of a theory, here, involves *conceptual* considerations which are independent of a theory's deductive forms. Another version of contextualism, one that has come into vogue quite recently, is to pay cognizance to the formal and empirical aspects of a theory while maintaining that a theory's true meaning depends on how these 'objective' features fit within an entire historical, social-political framework or discipline which was created and sustained by a community of scientists. Some social-political contextualists have even gone so far as to attempt to reduce the very meaning of the logical and objective features of theory in social-political terms, which would make them the paragon nemesis of the positivist and formalist camps!

One of the many tasks before us, then, is to examine critically the rival approaches to theory sketched above, with hopes of formulating a theory about theories which retains whatever insights each of the above has to offer but casts aside what is false or misleading. It is my contention that this can be achieved, to a great extent, if we adopt a stance which treats theories primarily as systems which depict the nature of things, not logical algorithms for the organisation and prediction of data. While a theory has definite logical, semantical, even social-political aspects, each one of these aspects being emphasised by one of the above schools of thought, we maintain that these features are actually highly dependent on the way the theory describes the nature of things, rendering the above views on the nature of theory, at most, mere approximations of what we believe theories really are. Much effort in the second part of the book is devoted to showing how a systematic depiction of the nature of things affects all the other aspects of theory. I will denominate this approach to the nature of theory *ontological contextualism*.

2. SCIENTIFIC REALISM VERSUS PHENOMENALISM

Showing that there is an inextricable connection between how a theory depicts nature and its other features constitutes a defence of a position in the philosophy of science known as *scientific realism*. Scientific realism maintains a commonsensical view that there is a world that exists independently of our perception of it, and that our theories inform us about the existence and nature of this realm, even if the things of which our theories are about are not directly observed or, in some cases, not even observable. The realist holds that the truth or success of a theory in some sense entails the existence of the theoretical entities mentioned by the theoretical terms in its language. For example, the success of atomic theory in terms of its ability to predict and explain a gigantic cluster of observable phenomena entails that atoms exist, even if we cannot see them. If we hold that theories contain such ontological commitments, one interesting task for the scientist and philosopher alike is to glean what kinds of entities a theory commits us to, whether they are particles, fields, mental states, social structures, and so on.

According to the realists, then, the exciting thing about theories is their ability to break through appearances, informing us about a new and, perhaps, 'mysterious' reality of, say, micro-particles, fields, etc., beyond our everyday experiences. Now the *phenomenalist view* takes the 'romance' out of theories by restricting their contents to nothing but the world of things that are immediately perceived. The belief that theories tell us about a world beyond appearance is replaced by a 'hard-nosed' empiricism. Theoretical entities are 'convenient myths' used to organise experience. The truth or success of a theory does not lie in its successful reference and description of non-perceivable entities but instead in the capacity of its theoretical terms to organise and predict experience. In fact, as we will soon learn, the logical positivists, who were phenomenalists *par excellence*, presented us with a definite programme for reducing the meaning of any scientific theoretical term to nothing but observable terms.

The dispute between scientific realism and phenomenalism raises several important questions about scientific practice. Can science function on an observational level alone, as the phenomenalist maintains, without any commitments whatsoever to the existence of unobservable theoretical entities? Can the meaning of theoretical terms be semantically reduced to those of observables without any loss of content? If theoretical entities are not perceived by the scientist, how can it be claimed that we know they exist? What is the connection between the ability of a theory to predict and organise data and the successful reference beyond appearances of its theoretical terms? It is my contention that answers to these questions will eventually favour scientific realism, that scientific realism is the only way to make sense of the workings of theory.

3. THE RELATIONSHIP BETWEEN THEORY AND OBSERVATION

The controversy between scientific realism and phenomenalism carries over into all kinds of claims and counterclaims on the nature of scientific observation, as well as the role theory plays in making a scientific observation. The phenomenalists and empiricists believe that any form of perception, whether it takes place in an ordinary or scientific context, is in a sense *given*, free of any theoretical presupposition. We perceive nothing but the 'bare' facts, so to speak. This is a well-established traditional view. It was most clearly stated in the writings of Hume and Berkeley, and, more recently, is to be found in 'sense datum' theory. The phenomenalist's opponents contend that there is much more to perception than the mere experiencing of sense data, that scientific observation is not incorrigibly given at all; for if we carefully examine what goes on even in making the most elementary measurement, we will soon discover that an observation is an achievement which contains many assumptions of a highly theoretical nature. For example, measuring the length of a table with a yardstick without worrying about the orientation of the table in space assumes, among other things, that space is isotropic and homogeneous.

Actually, each of the above stances on observation seems to be true and false at the same time! This can be put in the form of a dilemma:

Either observation is dependent or independent of a theory.

If it is dependent, how can observation be used to confirm our theories without begging the question?

On the other hand, if observation is independent of theory, how can we possibly account for the very obvious fact that our theories influence or determine what we claim to observe?

The relationship between theory and observation must be delineated in such a way as to resolve the above dilemma. Is there an enormous difference between what a cat sees while watching a mouse move across the floor and what a scientist sees when observing tracks in a Wilson cloud chamber? If there is a great difference in what they see, as many contextualists maintain, and if the difference lies in the fact that theory does not at all enter into the former case while it is ubiquitous in the latter, it behooves us to ask where and in what way does theory enter into the latter cases.

4. THE RELATIONSHIP BETWEEN THEORY AND THEORY

How to describe the relationship among theories rigorously has been a problem that has plagued philosophers for centuries. Even now, the philosophical literature is packed with a multiplicity of views on the matter, opinions which no doubt have been presented in an atmosphere of heavy controversy, with little or no unanimity in sight. Why should this be a problem? At first, there seems to be no problem at all. We simply have one world with a multiplicity of successful theories which describe it. For example, consider a student in a classroom who raises his hand to ask a question. We could apply a variety of theories to such an event. It could be handled sociologically, psychologically, physiologically, chemically, even in terms of microphysics. Now we ask: what is the (logical) relation between these theories which have a common subject matter?

Several answers to this question have been proffered. The stance taken by the logical empiricists has been that of reduction: the various theories form a partial ordering of levels of description, each level being more microscopic in description than its immediate predecessor. According to the logical empiricists, these levels are related by deduction, i.e., the results and laws of the macroscopic science being derivable from the laws and assumptions of the more fundamental, microscopic science. One of the major objectives of logical empiricism was the unity of the sciences which, they believed, could be accomplished by means of reduction. As they saw it, all scientific descriptions and accounts of events could be reduced to microphysics.

As tidy as this picture of reduction may have appeared, it has since turned out that there are insurmountable difficulties in using traditional logical techniques to capture the relationship between the reducing and reduced systems or levels of description. We shall go over this in detail in Chapter 8, but it appears that the precise relationship between the reducing and reduced science is a highly elusive one, even for non-controversial cases such as reducing macroscopic objects – e.g., tables and chairs – to atomic lattices. I am afraid, then, that until the formal relationship between the reduced and reducing system is made unquestionably clear, those scientists and philosophers who have faith that someday all macroscopic sciences will be reduced to microphysics are confronted with the problem that they do not really know what theory reduction *is*. So, if anything, we seek to find a description of that relationship in such a way that will do justice to actual clear-cut cases of reduction in the sciences. As yet, this has not been done.

On the other hand, if theories are not related by reduction, what are the alternatives? Well, dualism appears to be one: each theory describes its own isolated 'universe' while the strongest possible relationship between these separate and distinct systems is that of correlated events. (There are other variations on this theme, but I will not bother with them here.) Another is that of emergence: although each macroscopic system is made up of microscopic objects there are properties or features that cannot be reduced to combinations of microscopic properties. For example, properties associated with living organisms cannot be reduced to combinations of inert matter, even if the former is composed of the latter. The relationship between the emergent properties and those of the lower level, then, is that of correlation, not reduction. Carrying this to an extreme, perhaps each scientific system is unique to the extent that it is futile or artificial to believe that we can find a blanket relationship among diverse sciences. Which of these views has the greatest merit? Better yet: How can we rationally decide which one does? It is interesting to note that the reductionist–emergencist dispute, which has been going on for centuries, has claimants on each side who insist that there is *empirical evidence* in their favour! One of the things we must do, then, is to arrive at a precise, logical characterisation of each of these positions in order to investigate the possibility of there being a way empirically or methodologically to settle these issues.

5. THE EPISTEMOLOGY OF THEORY

Epistemology or theory of knowledge deals with questions concerning standards of knowledge and rational belief. We want to arrive at criteria which, for example, enable us to distinguish those circumstances in which one actually knows that something is the case from occasions where one only believes it to be the case and just happens to be right. We also seek criteria which tell us when a set of data constitutes solid evidence for believing that something is the case as opposed to, say, coincidental or circumstantial data. There are several other related questions which arise in the course of epistemological investigations, but the major ones which will involve us here concern criteria of 'theory satisfaction'. Of course, the so-called objectivity of scientific method is at stake in answering the above questions in the context of scientific observation and experimentation.

As it was for delineating the exact nature of the relationship between different theories, philosophers of science have also had a very difficult time developing criteria for the confirmation and refutation of a theory which does justice to the actual complexities of scientific practice. What we are looking for is a model of theory confirmation and refutation, one which tells us when it is rational to support or abandon a theory on the basis of data, observation or something else. Several such models have been proffered by philosophers of various persuasions but, I believe, without much success. Many of them, in fact, leave us with 'paradoxical' results such that our theories are never *really* confirmed. But is the general theory of relativity a confirmed theory and is Aristotle's physics known to be false? Some philosophers of science have even gone so far as to deny us access to the categories of 'truth' and 'knowledge' when it comes to fundamental scientific theories, something that would be quite a shock to many scientists who believe that science is the paradigm of truth and knowledge.

It seems that many questions surround these supposed criteria of theory acceptance or rejection. Is the process of theory confirmation disinterested and value free, or are there psychological, social and political forces at work behind a facade of objectivity? When theories are readjusted in the face of contrary or anomalous experimental results, are the 'rules' of change objective or do they entail values as well? There are those who believe that theory confirmation or refutation cannot be a purely objective affair, that judgements such as these are intertwined with the values of the scientific community. This contention has been motivated, in part, through the realisation by many philosophers that data, in themselves, do not seem sufficient to entail the confirmation or refutation of a theory. So they naturally conclude that other factors enter into the confirmation–refutation processes that carry us from collecting raw data to full-blown acceptance or rejection.

The above view that non-objective beliefs always enter into scientific methodology, a position which is diametrically opposed to that of the logical empiricists but much in vogue in philosophical and scientific circles alike, leads to the spectre of relativism in the sciences. If our methods of verification reflect values, metaphysical beliefs and social-political pressures at work in the scientific community, does it not follow that our cannons of testing are internal to that particular community? If the answer to this is 'yes!', how can we objectively compare rival, fundamental theories other than by pitting one set of values against another? But this makes scientific practice, which was originally envisaged as the epitome of objectivity, look more like an evangelising activity where persuasion has replaced proof or demonstration. Many philosophers of the social-anthropological school of science welcome such a result. The same applies to the traditional view of scientific progress as the accumulation of knowledge. Many feel that progress is another measure of scientific objectivity in that the replacement of one science with a successor means that we are getting better theories, that we are getting closer and closer to the truth. This opinion has been subject to a series of attacks quite recently, calling out for new, more social-political models of scientific change.

While I must disagree with the sociological attack on scientific objectivity, this trend may have done us a service by successfully putting an end to the rather naive versions of confirmation, scientific progress and theory comparison which have been put forth by the logical empiricists and formalists. We simply cannot confirm our theories, demonstrate that one theory is superior to another of its kind by 'pointing to the facts' alone. Things just do not work that way in the sciences, nor do they in everyday life. What we really need are models of theory confirmation and theory comparison that do justice to the complexities found in actual scientific practice but where we rightly feel at home in calling these procedures objective.

Each of the approaches to theory mentioned in section 1, for the most part, supplies answers to the above questions. Let us now critically examine what they have to say about these issues, beginning with logical positivism.

2   Logical positivism

1. INTRODUCTION

One origin of logical positivism can be traced to the writings of the French social philosopher, Auguste Comte. Comte maintained that scientific theories underwent a three-stage evolution: theological, metaphysical and positive. In the theological stage, theories contain anthropomorphic elements, animating objects or appealing to the will of a deity in order to explain events. Although the metaphysical stage does not appeal to teleological factors, theories still explain by citing ultimate forces or causes underlying observed events. Only until theories are purged of the above theological and metaphysical elements by restriction of their contents to 'relations of succession and resemblance', i.e., to laws or relations among phenomena, can they reach their most advanced, positive stage. Scientific theories, then, were nothing but collections of generalisations about phenomena.

Logical positivism maintained two major objectives: to remove the pretensions of the idealist philosophers who believed they had a unique metaphysical, supra-access to truth, one not available to the scientist; in so doing, the proper boundaries of philosophy and science are made clear in order to ensure that their respective enterprises will not infringe on each other, as they have in the past and with most confusing results. The logical positivist used powerful mathematical and logical systems as the major tool of analysis in order to achieve the above two objectives. Like Comte, they wanted to purge teleological and metaphysical elements from the content of scientific theories, and from philosophy as well. In fact, according to these philosophers, metaphysics was to be stripped of any legitimacy whatsoever, no matter in what context it may have occurred. This is where the logical positivists separated from other philosophers who used logic and mathematics as a major means of doing philosophy, in particular, their 'cousins', the logical atomists.[1]

Before we go into the tactics of the above move to purge theories of metaphysics, let us briefly go into what metaphysical claims are about. Those who did metaphysics were concerned with arriving at truths concerning the nature of things and those principles which served as the ultimate foundations of scientific systems. Since the sciences were supposedly dependent on metaphysical principles, the latter were not subject to empirical support but were *a priori* truths, i.e., claims whose truths were established independently of experience. They included claims such as 'Every event has a cause', 'The effect resembles the cause', 'The future resembles the past', 'Space and time are absolute', 'Space must be Euclidean', etc., that had a ring of necessity by going beyond the evidence of the senses although they seemed to be essential for the existence of science itself. Notice how some of these claims were about lawlike necessity. This prompted the positivist to take up David Hume's battle against the doctrine of causal necessity or necessary connections. This view maintained that changes in nature were brought about by necessary connections among objects in nature, i.e., relations that depended on the inherent properties (causes or powers) of the objects involved and that the nature of these relations was such that once we knew the inherent properties of things, we could actually calculate the course of events in an *a priori* fashion. For example, if two rigid gears were meshed and one turned clockwise, we would know *a priori* that the other *must* turn counterclockwise. Nothing else was conceivable. Naturally, such a view of causation was repugnant to those of an empiricist bent. The reaction of the logical positivist was to replace necessary connections with a much weaker relationship among events, that of *correlations*.[2] Laws of nature, then, did not refer to necessary connections or powers but instead to generalisations concerning the ordering of events. Correlations are the strongest possible relation expressible in a scientific theory and in philosophy.

But there is an aspect of doing metaphysics which the logical positivist found to be equally abhorrent, even if it lacked the *a priori* features of metaphysical truths that were claimed to support scientific theories. I said earlier that metaphysics was concerned, in part, with the nature of things, that is, with what kinds of entities there are in the universe (as a whole). These concerns are identified with the study of ontology, and many scientists and philosophers believed that scientific theories held answers to these questions; in fact, for many, it was the major motivation for doing science. Yet, the legitimacy of even this enterprise was to be denied by the logical positivist. There are some philosophers and scientists who believe that the truth of scientific theories has ontological implications, that their truth often entails the existence of unobserved and even unobservable entities. (I will often refer to this belief as 'scientific realism'.) So, not only are the *a priori* claims of metaphysics to be rejected as mystical and irrational, so are any ontological implications about unobserved entities from true but empirically supported theories. The only thing left for a science to do is to present a catalogue of experience.

According to the positivist, then, a pure scientific theory was an algorithm for the codification of the scientist's experiences. Contrary to what may have been believed by philosophers and scientists, theories do not serve as guides to a realm beyond experience.[3] I have said above that they made use of the tools of logic and mathematics in order to arrive at this result. It is to this technique of analysis of scientific theory that I will now turn.

2. POSITIVIST ANALYSIS OF SCIENTIFIC THEORY

As should be clear by now, metaphysical claims are part of neither science nor philosophy. If philosophy is not concerned with metaphysics, what is left for the philosopher to do? The answer, according to the logical positivist, is *to analyse the language of science*: that is, semantics is the legitimate domain of philosophers; anything else is to be avoided. How is the philosopher to analyse the language of the scientist? By systematically transforming the syntax of scientific propositions in terms of symbolic logic or mathematics along with the task of reducing the meanings of the theoretical terms which occur in these propositions to those of observables.[4] This is how scientific systems are reduced to the codifications of nothing but observables and how science is rendered purely empirical.

We can see how operationalism, i.e., construing the meanings of (theoretical) terms as a function of a battery of tests, measurements or observations, goes hand in hand with logical positivism. What may have been taken to be the *ascription* of a theoretical property to a system in expressions such as 'These atoms are negatively charged' or 'Salt is soluble in water' turns out to be some sort of abbreviation for a set of correlations among observations or test behaviour. In the 'Salt is soluble' case, what is really meant, in part, is 'If we were to place salt in water, it would dissolve.' In this way, the meanings of theoretical expressions are reduced to configurations of test sentences.

Let us see how the above meaning reduction works in detail. As I stated above, this was to be achieved by means of symbolic logic or mathematics. In the former case:

'X has theoretical property, T' means if X is placed under test conditions, C, then such a test will yield observable results O.

For example, 'X is magnetic' means that if a compass is placed in X's vicinity, the needle will be deflected; or, 'X is hungry' means if given the opportunity to eat, X will do so, and so on. The same move can be made mathematically. Newton's second law, $F = m\ a$ (force equals mass times acceleration), contains two theoretical terms 'force' and 'mass'. In *The Science of Mechanics*, Ernst Mach performed a two-stage reduction.[5] First, he converted $F = m\ a$ from a law to a definition of 'force', i.e., 'force' now means the product of mass and acceleration. Now 'mass' is reduced to observables in this way. The third law states that for every action there is an equal but opposite reaction, $F_{12} = -F_{21}$ (where 'F_{12}' denotes the force body one exerts on two). Thus $m_1a_1 = -m_2a_2$ or $m_1/m_2 = -a_2/a_1$. So if two bodies bounce off each other, their reaction accelerations will depend on their respective masses. The greater a body's resistance to motion, the less it will accelerate after colliding with another body. Now all we have to do, according to Mach, is to choose a standard mass and use it to scale masses of all other bodies by observing their reaction accelerations *vis-à-vis* our paradigm object. When such an analysis is systematically applied to all theoretical terms in mechanics, the science of mechanics turns out to be nothing but a group of laws (and definitions) which relate the movements of bodies. The sole residues of analysis, then, are observables.

By the way, there was some debate among the logical positivists as to what observable terms referred. Some philosophers such as Carnap felt that they denoted the experiences of the scientist. They might include, for example, a scientist's experience of a red object in a laboratory or experiencing a clicking sound made by a Geiger counter. The statements which referred to these experiences were called 'protocol statements' while the language of observation was the protocol language. Protocol statements such as 'here, now, blue' or 'there red' were to be the fundamental units of scientific language as well as the empirical basis of a scientific theory since their truth needed no further justification; they were, in effect, incorrigibly held.[6] But experiences are not the same as physical objects or events of which they were supposedly about. Other logical positivists made physical objects and events the basic unit of observation, reducing theoretical propositions to logical functions of sentences that referred instead to these public objects. This was known as physicalism.

Another version of the reduction of theories to relations among observables can be found in P. W. Bridgman's *A Sophisticate's Primer of Relativity*.[7] Bridgman felt that the special theory of relativity represented the culmination of logical positivism, something Einstein fervently denied, as we will see later on. The reason why Bridgman believed this was that with the special theory of relativity, theoretical statements such as '*A* is simultaneous with *B*', 'momentum is conserved', etc., were quite amenable to the above type of meaning reduction, this time, in terms of standard clocks and measuring rods. Again, 'length', 'velocity', 'momentum', etc., were terms that did not denote properties of things. How could they if we get *different amounts* of each when one and the same object is observed under various inertial frames of reference? We are better off meaning by these terms how various measuring instruments behave under a variety of experimental conditions which are relativised to the observer's inertial frame. Not only does the special theory of relativity show us how to do this, it also tells us how we can relate the behaviour of instruments among different inertial frames. This elegant picture of the relativised behaviour of clocks and rods was captivating indeed! It was to supply much ammunition for attacks against past metaphysical claims in mechanics. As empirically powerful as Newtonian mechanics was, it had one pernicious vestige of metaphysics: absolute space and time. Newton felt that if his laws of mechanics were true, then space and time were absolute, i.e., bodily motion was the change of location in absolute space with respect to absolute time. In other words, Newton believed that his mechanics not only dealt with bodies that consisted of mass points which obeyed his laws, but also that all of this took place with another *entity*, one very unlike physical objects and whose geometrical properties were independent of the objects contained within. Moreover, Newton readily admitted that absolute space and time could not be directly observed, that their existence could only be inferred indirectly from certain rotational motions of bodies. It can be seen how such a metaphysical doctrine coming from such a highly successful scientific theory was a source of embarrassment for the logical positivists. We can imagine the sigh of relief they felt when Mach and Bridgman 'showed' how motions characterised as change of absolute place with respect to absolute time could be done away with in terms of the *relative* motions between bodies. The latter can be readily observed. This is why they felt that the special theory of relativity delivered the final blow to metaphysics in the science of mechanics.

Another reduction occurs in terms of explanation, especially causal explanations. As mentioned above, many philosophers believed that the Newtonian paradigm of science made forces the ultimate sources of change. Theories, then, explained by appealing to forces or mechanical connections in nature, a view that was subject to vicious and, many believed, devastatingly successful attacks by Berkeley and Hume. As it was for other theoretical terms, the notion of causal nexus is to be replaced by observables. According to Hume's analysis of causation, the only objective features of the relationship were temporal succession, contiguity and regularity among events.[8] I will focus on Hume's analysis in the next chapter but an example of such an analysis would look like this: 'Lightning causes thunder' means whenever lightning occurs, thunder immediately follows in the proximity of the lightning. So theories explain the occurrence of individual events, not by citing objects that, in themselves, have powers or properties 'built into them' in such a way as to bring about the event in question but because certain types of events are simply correlated with others – again, something that can be directly observed. Explaining events is a matter of seeing how their occurrences fit in with the laws of nature. The laws of nature are not metaphysical connections among events which are brought about by some sort of 'metaphysical glue', but are, instead, simply *descriptions* of how these events are correlated.

The logical positivists thus arrived at what has often been referred to as an inductivist conception of scientific theory. Theories are inductively built up out of observables alone while the logic of the theory describes how they are structured. The truth tables or logical matrices of the logicians are simply filled in by the accumulation of scientific data. Mach spoke of scientific theories as the generalisation of experience. This means that, contrary to the realist interpretation, theories are simply epistemological systems, i.e., algorithmic devices for organising and predicting data. Any role of theory beyond this is 'metaphysical' in a pejorative sense, a role certainly not worthy of being called scientific.

3. THE VERIFICATION PRINCIPLE OF MEANING

Sometimes, in the course of a conversation with a scientist who might be of the 'hard-nosed' school of experimental science, one might hear phrases such as 'If I can't measure it, it doesn't exist' or 'If you don't tell me how I can test your hypothesis, I haven't the least idea of what you mean by it'. Whether our experimental scientist realises it or not, such restrictions on what we can claim to exist or mean by an hypothesis presuppose the verification theory of meaning.

The verification principle of meaning was the major weapon employed by the logical positivist against the metaphysicians. As the principle says, there is an inextricable connection between the meaning of a sentence and its verification.[9] Since, by its very nature, a metaphysical proposition is not empirically verifiable, it is not even worthy of being judged true or false: metaphysical claims are meaningless. For example, how can theologians claim that God exists, or quantum physicists that electrons spin, if the very expressions used to make these assertions can be shown to lack meaning? I can put this another way in terms of the logical positivist's view of language, a view that perhaps can be traced back to Hume's distinction between relations of ideas and matters of fact.[10] According to Hume, knowledge claims fit one of two categories: they are either *relations of ideas*, 'discoverable by the mere operation of thought', such as $2 + 2 = 4$ or the Pythagorean theorem; or, they are *matters of fact*, such as 'The sun will rise tomorrow', which are based upon experience. While the former are subject to demonstration in that to deny each would lead to a contradiction, the latter are clearly not, since the denial of any matter of fact never implies a contradiction. With the development of powerful systems of symbolic logic, those philosophers who believed that logic captured the essence of language, in particular, the language of science, maintained a similar distinction. Sentences were either tautologies or analytically true, i.e., true no matter what the state of affairs may be; contingent or synthetic, i.e., true or false, depending on the actual state of affairs; or, they were logically false or contradictions, i.e., they were false no matter what the state of affairs may be. For example, 'It is raining or it is not raining' fits the first classification, 'It is raining' and 'It is not raining' are exemples of the second while 'It is raining *and* it is not raining', the third. Any proposition must fit one of the above three classifications.[11] Then what of the propositions of metaphysics? They are surely neither contingent nor logically false. But if they are logically true statements they are uninteresting in the same way that the tautology, 'It is raining or it is not raining' really says nothing about the weather. This is why the logical positivist felt that metaphysical propositions had no significance, that the bulk of propositions of interest were the contingent claims of science. The analytic–synthetic distinction, then, was one important source of the connection between meaning and verification.

Now there are two versions of the principle, a weak and a strong version.[12]

Weak Version: A sentence is meaningful if and only if it is verifiable.

This empiricist restriction on meaningfulness strips away the significance of propositions such as 'God exists', 'Every event has a cause', and other metaphysical claims. It also served to point out hypotheses that are, in fact, pseudo-scientific. One famous example of this can be found in the often debated case of the nocturnal doubling hypothesis. Suppose while we were asleep one night everything in the universe doubled in length, width and height. Could we verify that such a process took place after we woke up? Many philosophers who have studied this question say 'No', and we can understand why from a few elementary considerations. For example, we can not use metre sticks to see if the doubling took place since they, too, have doubled in length. Likewise, even though the period of a pendulum should end up being a quarter of what it was originally because its length is halved, this is compensated for by a decrease of gravity. So the reasoning goes. However, according to the weak version of the verification principle of meaning, because nocturnal doubling can not be detected, it is *not* really a hypothesis at all. In fact, some philosophers maintain that length is a relational property, and that no one object can be said meaningfully to change length at all if all other objects compare lengthwise to it in exactly the same way. This relational view of a property leads to the strong version of the verification principle of meaning.

Strong Version: The meaning of a sentence *is* its means of verification.

In other words, with the strong version, instead of a condition placed upon the meaningfulness of a sentence, we now have at our disposal a device for *analysing the meaning of a sentence*. It can be understood how the strong version fits in with operationalism. What would it mean, for instance, for an object to be three metres long? Well, how could we tell if it were, in fact, three metres long? We could take a metre stick and see if it can be placed along the length of the object three times, endpoint to endpoint. This is how we could tell and, by the strong version, this is what it would mean to say the object in question is three metres in length. Likewise, to say an object is red might mean that its colour matches the colour of the standard or paradigm red which can be found on a colour chart. But this is also how we verify that it is red.

The use of standards or paradigm objects in the course of ascription and measurement brings out another important feature of logical positivism, namely, conventionalism. Conventionalism is the view that if two theories are equally compatible with the observational facts, although they may appear to be rival theories, they are not really incompatible at all. In order to illustrate this, let us consider a classic case found in Reichenbach's *The Philosophy of Space and Time*.[13] Suppose we have two 'rival' theories about the nature of space and conjoin them with hypotheses about how measuring rods behave as they are transported from one region of space to another. The first theory, T_1, says that the geometry of space is Euclidean while the length of a rigid rod remains unaffected under transport. Theory two, T_2, says that space is non-Euclidean but measuring rods are systematically subjected to expansion forces as they are moved about various regions of space. Now we can 'juggle' the expansion force field in T_2 in such a way as to yield exactly the same observations as T_1.

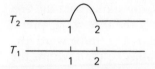

In this case, our observations would consist of distance measurements between points in space by 'filling up' the distance between points. This is done by laying down rigid rods, endpoint to endpoint; we then count the number of rods it takes to fill in a given space. T_2 says that space is curved. Normally it should take more unit rods to cover the 'hump', d_{12} in T_2 but if a 'force field' in curved space causes rigid rods to expand inversely in proportion to the distance from the midpoint of 12, then the number of rods it takes to cover d_{12} in T_2 is the same as that for d_{12} in T_1. So even if T_1 and T_2 respectively denote flat and curved spaces, their observation results are exactly the same, as it was in the case of the nocturnal doubling hypothesis.

If T_1 and T_2 are equally compatible with the observational facts, then according to the verification principle, T_1 and T_2 are *not* different theories. *They are simply different ways of presenting the same body of facts.* If so, why choose one theory over the other? Reichenbach's answer is that there is no objective reason to do so, that such a choice is really conventional in nature: it is simply a matter of deciding what we will *call* the 'same length' or how we define congruence. In other words, once we choose our standard (or paradigm) measuring rod,[14] we then stipulate whether or not its length is preserved under transportation. If we assert that its length is so preserved under transport, this amounts to accepting the Euclidean view of space, while asserting that measuring rods are subjected to expansionist forces which operate independently of the physical make-up of each rod is to accept that space is non-Euclidean. This is a conventional move, however, and how we make it is tantamount to choosing T_1 or T_2. Contrast this position with the realist interpretation of theories mentioned above. The realist maintains that T_1 and T_2 are inconsistent theories, that the choice between T_1 and T_2 is not a matter of convention but one of *truth*. Both sides of the dispute will disagree on the status of the verification principle.

A similar conventionalism can be found in Carnap's distinction between 'internal' and 'external' questions.[15] Relative to a given theory, we can ask whether or not there are, say, electrons. Such a question, according to Carnap, is really a question about the *term* 'electron', namely, how it functions within the theory. If, after a semantical analysis, we find that it functions as a 'thing word' (its role in the language is to designate an object), then within the theoretical framework there are electrons. However, it is an entirely different question if we should accept the reality of the framework itself. It is one thing to say that there are subconscious intentions according to Freudian psychology but quite another to say that Freudian theory is true. How do we establish that? According to Carnap, questions about the reality of the theoretical framework are *external* questions which are open only to *practical* considerations. So, whether or not there are subconscious intentions is settled by analysing the language of Freudian theory. Whether we accept the reality of Freudian theory is a conventional matter, i.e., depending on whether or not we choose to accept the language of the framework itself. Again, we have the contrast with the realist who insists that the reality of the framework is not a matter of choice but one of truth.

Logical positivism has left us with this picture of science and philosophy. We have seen that the verification principle of meaning, symbolic logic and mathematics have been combined to analyse scientific theories with the result that a theory simply describes and organises factual data. The content of a theory is completely exhausted by the facts it encompasses or by its corresponding body of test sentences. Anything else goes beyond the evidence of the senses and is to be thrown out as metaphysics. The scientist is to seek out new experiences and new generalisations of experience while the philosopher simply seeks to illuminate the syntax and semantics of the scientist's theories. The respective tasks of science and philosophy have been clearly defined.

4. VARIOUS ATTACKS ON LOGICAL POSITIVISM

The cornerstone of logical positivism is the verification principle of meaning. One wonders what form logical positivism could possibly take without some version of this principle or at least the philosophy of language described in the last section. An irony concerning the verification principle is that its own status becomes immediately open to question. After all, how does one verify its truth? If the logical positivist claims that its justification is really an external question, why should we choose to accept this framework (rules pertaining to meaning) in light of the fact that some of us may feel that such a principle places restrictions on scientific theories that are much too severe? Not only that, the very status of scientific laws is equally open to question, for how can we, at this present moment, verify the truth of a law of nature when its scope clearly goes beyond that of the past and present into the future?

In a very important way, the verification principle presupposes that the meaning of any term which stands for an observable is independent of theory; otherwise, the meaning of a theoretical term could not be defined by observables. This point has been hotly contested by contextualists such as Feyerabend, Hanson and Kuhn. They contend that, to the contrary, the meaning of an observable term is a direct function of the theory in which it is embedded. For example, the meaning of the word 'distance' is really wrapped up in our theories of geometry and because of this, its meaning will vary as we go from one type of geometry to another. 'Distance' in flat, Euclidean geometry is not synonymous with the term as it is used in non-Euclidean frameworks. In other words, the meaning of observable terms more or less cascades from basic theoretical assumptions. Theory, then, supplies a context or a set of background conditions for the meanings of its terms. So, the contextualist presupposes a theory of measuring which is completely opposite that of the positivist's assumption. The details of specific versions of the view that meaning is theory dependent will be examined in Chapters 4 and 5.

Not only must the positivist make the dubious assumption that the meanings of observational terms are independent of theoretical ones, he is also committed to the supposition that scientific observation and measurement are given independently of theory. This leads to another line of attack against logical positivists, one which can be found in the writings of Hanson and many other philosophers of science. It concerns the supposed independence and given nature of observables or data. When the logical positivist maintains that theories can be reduced to configurations of observables, he is assuming that the identity conditions for observables are independent of theory, i.e., determining what we observe is in some sense given regardless of theory. This is precisely what Hanson denies when he argues that observations are 'theory laden', that scientific observation is not independent of theory. For this reason, many philosophers have argued that the operational programmes of the logical positivists are really a charade, for most of them end up smuggling in theoretical terms in order for their operational definitions to work. These philosophers do not believe that a science completely free of theoretical terms can possibly be developed. But, if theoretical terms can not be reduced to observables and if determining what we observe requires theory, how can it be said that theories are reducible to configurations of observables? Observables are just the tip of the iceberg.

Even the positivist's seemingly innocuous distinction between observable and theoretical terms is questionable. Achinstein does just this in his *Concepts of Science*.[16] According to the logical positivist and the verification principle, a term which refers to an observable must differ in meaning from one which denotes a theoretical entity. Yet, as Achinstein points out, it is highly intuitive that Maxwell's use of 'velocity' in the velocity of molecules, which is *not* an observable, is 'exactly the same were molecules much larger and therefore visible'. To cite another, more commonsensical case, he notes that 'I am not using the word "tomato" in a new way, nor am I telling a fairy tale, when I claim that the tomato in the box is not observable (because the box is closed).'

One result of this dispute between the positivists and the contextualists, something which Achinstein carefully shows, is that two, extreme views on the relationship between the meaning of observable terms and theory have evolved. On the one hand, the positivist holds the meaning of an observable to be completely independent of theory while, on the other hand, the contextualist maintains that its meaning highly depends on the theory in which it occurs. This issue will come up again in Chapter 5 but what we really need in order to settle this dispute is a set of criteria or a model of how both theoretical and observable terms get their meanings. According to Achinstein, the correct stance to take is somewhere in between, that both sides of the dispute falsely presuppose that there exists a clear-cut distinction between theoretical and purely observable terms in the first place. A correct way to go about this, then, would be to take each term in a case-by-case study and examine to what extent theory contributes semantically relevant conditions for its meaning and reference.[17] But such a technique requires that it is possible for the meaning of some terms to be independent of theory while the meaning of others may be highly theory dependent, with an entire spectrum of dependence in between. As it stands, neither side of the dispute has provided adequate backing for so important an assumption about the relationship between theory and the meanings of words that occur within.

Other criticisms of logical positivism question the verification principle of meaning itself, by putting forth counterexamples. What would a counterexample against the strong version be like? It would simply be a matter of finding a case where the meaning of an expression or a sentence can not reasonably be identified with the techniques used to verify its truth. As several critics of the verification principle have pointed out, such cases are readily found when it comes to scientific claims about past events. This is because, they maintain, the past can not be directly observed. How can the meaning of a sentence about the past be its means of verification when any verification procedure or test takes place in the present? This entails that statements about the past are reduced to tests about the present![18] For example, the meaning of the sentence 'A meteor which hit the Earth 60 million years ago caused the extinction of dinosaurs' is reduced in *meaning* to statements about the amount of iridium found in a particular layer of the Earth's crust, present-day observations of fossil records, carbon 14 dating tests, etc. The outcome is that it looks like a straightforward application of the strong version of the verification principle to statements about the past leading to this highly unintuitive and paradoxical result that they are not really at all about the past but depict measurements which take place in the present.

The strong version also has a problem concerning the sameness or identity of observables. It seems to be incorrigibly true that it is possible to measure one and the same observable many different ways. While we may use a thermometer to measure the temperature of a gas, we could also measure that very same property with a thermocouple or a bimetallic device. In fact, there should be no reason why temperature can be measured an indefinite number of ways. If so, does the strong version entail that there are several 'temperatures' of a gas corresponding to the many different ways we verify that a gas has a specific temperature? If we have several ways to measure duration (with different types of clocks) does this mean that there are several times which are measured? If the positivist wishes to avoid a plethora of observables, he must come up with criteria which tell us which cluster of instrumental readings go with a specific property being measured. Not only that, the critics of logical positivism will insist that such criteria can not work independently of theory, something which is itself incompatible with the verification principle.

Even the weak version of the verification principle can be questioned, and we can use the nature of scientific claims about the past, again, to generate counterexamples. The weak version states that a sentence is meaningful if and only if it can, in principle, be verified. This entails that any meaningful statement about the past must be verifiable. But the only way we can determine the past state of a system is to observe its behaviour in the present. So, for example, we can use the present position of the planets in conjunction with the laws of mechanics to determine their location 1000 years ago. Now, what if the present behaviour of a system does not supply us with sufficient information to distinguish between various states it may have been in the past? Thus, can we not verify the exact nature of its past? The weak version can not tolerate such a possibility but it remains to be seen how it can be ruled out in light of the fact that cases like these can be found in contemporary physics. Placing rigour aside, one formal result of quantum theory is that it is possible to prepare two differing beams of electrons in such a way that if we were subsequently to run every conceivable experiment on them, we could not, *in principle*, tell which one was in what state. Their present and future experimental behaviour would be indistinguishable. That their past states differed while their present behaviour is indistinguishable is a direct result of quantum theoretical postulates. So, one wonders how adherents of the weak version could reasonably question such a possibility except to challenge the very foundations of one of today's most powerful scientific systems.

As usual in philosophy, there is a way out for the logical positivist, one which involves a reformulation of the verification principle of meaning. Instead of speaking of verification in terms of making observations in the present, verification includes a claim about the past, namely, what *would have been* observed about past systems *were* certain experiments performed. Thus, if we were to observe how the systems behaved back then, in the process of preparing them we would have noticed the difference. Such claims about *what we would have observed when* are known as contrary-to-fact conditionals concerning observations. Their form will be discussed in the next chapter. What this escape route involves is that the verification principle of meaning be restated where verification includes the above contrary-to-fact conditionals pertaining to observation. Even so, it does not end here, for we shall soon see that the critics of logical positivism will retort that the truth of these counterfactual claims especially can not possibly be established without resorting to theory and its accompanying ontology of the very theoretical entities whose access we are to be denied. The positivist can not have it both ways by making use of the very entities he wishes to discard. This very point will be elaborated in Chapter 9.

What the above cases do bring out is this commonsensical and often mentioned point about verificationism: it would appear that an elucidation of the meaning of a theory is most often *logically prior* to determining how it can be verified. Perhaps logical positivism has been a bit overzealous in its attempt to purge non-empirical, dogmatic claims from science by motivating an unwarranted connection between meaning and verification (or confirmation). We have seen that the positivist has failed to provide us with a theory of meaning which justifies such a strict tie-up between the two. That scientists should support their claims by verification procedures in order to avoid the pitfalls of speculation is something which I most wholeheartedly endorse. However, it is another matter to claim, as the positivist does, that any kind of a metaphysical claim, especially those about the nature of things, is not subject to verification and, hence, it is not responsible to make such a claim. Of this, I most emphatically disagree, and much of the text will be devoted to showing how the various claims scientists make about unobservable entities such as electrons, photons, quarks, etc., are ontological statements which are quite open to verification.

In relation to the above, although I agree with the critics of logical positivism that there is something amiss with the theory of meaning which serves as the foundation of their school of thought, there has most recently occurred a new line of àttack which I intend to pursue. Instead of showing that the positivist can not successfully reduce the meaning of theoretical terms to configurations of observables, this latest move embraces the positivist doctrine that the best theory is the one which is well-confirmed over its rivals. Then it is argued that theories which *refer* to entities underlying observables but can not be observed themselves are better confirmed than those positivistic theories which refer to nothing but observables. This line of argumentation will be developed in Chapters 6 and 7.

Finally, any good analysis of a scientific theory must somehow account for its explanatory power. Even if, as the positivists believe, theories are merely epistemological devices, they nevertheless bring about understanding. The question is, does the resultant analysis of a scientific theory presented by logical positivism do justice to the fact that theories generate powerful explanations as well as predictions? If the logical positivist denies this, he must show that theories do not really explain. On the other hand, if he accepts this connection between the nature of theories and their explanatory power, he must show how organised systems of data yield explanations. Yet, there are philosophers of science who argue that systems of observables which just 'repeat the facts' have no explanatory power whatsoever, that logical positivism can not possibly account for how theories generate understanding if the ontology of a theory is restricted to observables. Of course, this issue will be covered throughout the text with an eye towards this criticism that positivistic analysis of theories does not do justice to scientific explanation and understanding. In fact, I will argue that their whole approach is highly amiss, and that instead of presenting an analysis of the *language* of science, the philosopher would be much better off examining the rich and various ontologies *expressed* in theories. But this is precisely what the logical positivists denied theories could do. In the ensuing chapters, I intend to show why they are mistaken.

NOTES AND REFERENCES

1. Two excellent sources on logical positivism and its relationship to other contemporary philosophical schools of thought are J. O. Urmson's *Philosophical Analysis* (London: Oxford University Press, 1956), and G. J.

Warnock's *English Philosophy since 1900* (London: Oxford University Press, 1958).

2. Cf. R. Carnap's *Philosophical Foundations of Physics* (New York: Basic Books, 1966) pp. 12, 189–93.
3. Warnock, op. cit., p. 137.
4. Urmson, op. cit., pp. 102–29.
5. Ernst Mach, *The Science of Mechanics* (Chicago: Open Court, 1960) pp. 264–71.
6. See Urmson, op. cit., pp. 119–22.
7. P. W. Bridgman, *A Sophisticate's Primer of Relativity* (Middletown: Wesleyan University Press, 1962).
8. Hume, *Treatise* (Oxford University Press, 1960) pp. 155–75; *Enquiries* (London: Oxford University Press, 1957) pp. 60–79.
9. See A. J. Ayer's, *Language, Truth* and *Logic* (New York: Dover Publications, 1936) pp. 5–15.
10. Hume, *Enquiries*, pp. 25–39.
11. R. Carnap, *Meaning and Necessity*, pp. 1–15; 222–9.
12. These versions should not be confused with A. J. Ayer's weak and strong sense of 'verifiable'. His distinction, op. cit., p. 9, deals with partial as opposed to conclusive verification.
13. Reichenbach, *The Philosophy of Space and Time* (New York: Dover Publications, 1958) pp. 1–19.
14. Reichenbach calls this a coordinative definition (op. cit., pp. 14–19) because it links an object with a concept in the same way that the standard metre in Paris is linked up with a unit of length.
15. Carnap, *Meaning and Necessity*, pp. 205–21.
16. Achinstein, *Concepts of Science* (Baltimore: Johns Hopkins, 1968) pp. 106–19.
17. Ibid., pp. 199–201.
18. Warnock, op. cit., p. 49.

3 The deductive–nomological approach to theory

1. FORMALISTIC APPROACHES TO THE PHILOSOPHY OF SCIENCE

Throughout this text I will be tracing and evaluating, as I go along, the evolution of the concept of a scientific theory, beginning with that portrayed by logical positivism and ending with Kuhn's historical approach to the philosophy of science. As we shall soon see, the link between the positivists and Hempel appears to be quite logical and continuous but, after that, the philosophy of science abruptly diverges into two radically opposing schools of thought, namely, the formalists on the one hand, and the contextualists on the other. The proponents of the formalistic approach to the philosophy of science have included Nagel, Salmon, Scheffler, Suppes *et al.*, while Hanson, Kuhn, Scriven, Toulmin *et al.*, have been counted among their contextualistic adversaries.

The formalists believe that the essential aspects of a scientific theory, including its explanatory power, can be captured by bringing out its logical structure. Theories, for the most part, are complicated and powerful deductive systems which yield predictions and explanations. Thus, if we are truly to appreciate and understand how theory works, we must go into the logical syntax of the theory's language. Notice how this approach presupposes a particular philosophy of language, namely, that the propositions of science have two fundamental aspects: logical syntax plus reference. Once we identify the logic of scientific propositions and the range of objects denoted by the non-syntactic terms which occur in these sentences, we have, in effect, the major information required for understanding the meaning of the theory's language. Any residue of meaning can be discarded as being psychological or pragmatic but of no real concern to the philosopher of science

32

since his role is really that of an applied logician. Very often, such non-logical aspects of a theory have been given the label 'vague' and not worthy of analysis. The contextualists, on the other hand, have wholeheartedly accepted Wittgenstein's criticisms of the above philosophy of language. This critique will be described in Chapter 4, but its upshot is that we can not really grasp the meaning of the sentences in a scientific theory simply by examining their logic and reference; that, on the contrary, there are extralogical and extrareferential elements in the language of theory that can not be so dismissed by the formalists as being unessentially pragmatic or psychological. These elements, instead, have more claim to being the essence of a theory than logical syntax plus reference. They range from the semantical aspects of the terms of the theory (e.g., Hanson's approach) especially their interrelations with other terms, to even the social-political situation in which these terms are used (as Kuhn would have it). Unless we go into these extraformalistic aspects of theory, the contextualist argues, any analysis of theory will be amiss. This difference in the approach to theory will be one of the major topics in the text and, in Chapter 10, a resolution of this dispute will be presented.

Probably the major spokesman of the formalist approach to scientific theory has been Hempel. Decades have passed since the formulation and development of the Deductive–Nomological model (henceforth, D–N) of explanation. For good or ill, this model has played a dominant role in the philosophy of science as the major tool of analysis of scientific theory. Over the years, it has been carefully scrutinised by almost every philosopher of science. In spite of the many attacks it has received and various proposals of new approaches to explanation, it must be admitted that the D–N model's elegance and force remain among philosophers and philosophically-minded scientists alike.

Although Hempel disavows logical positivism, he does share with the logical positivists the view that the power of a theory lies in its logical structure which, in turn, gives it the ability to organise a vast array of data. Unlike the logical positivists, however, Hempel argues that the very use of symbolic logic to analyse theories entails that theories are not reducible to observables, and that the meanings of theoretical terms are not reducible to observation terms.[1] For Hempel, theories are highly complex systems of inference. And notice that because we know of at least two different kinds of inference, namely, deductive and inductive, there are two distinct types of theories and explanations.[2]

In any case, if Hempel is correct in his contention that the logical syntax of a theory accounts for its ability to predict and explain a given phenomenon we are provided with a nice recipe for analysing a theory. Translate the language of the theory into symbolic logic; sort out the various types of propositions according to their logic; and, by using the D–N model as a guide, identify the set of predictions and explanations generated by the theory.[3] In other words, we can arrive at the essence of the theory independently of any semantical, psychological or social-political contexts in which the theory may occur – something Hanson and Kuhn will most emphatically deny.

According to Hempel's analysis, an explanation of a scientific phenomenon is a set of statements (often called the *explanans*) which is logically related to a description of the phenomenon (called the *explanandum*). If the explanation of the phenomenon is a good one, then these four conditions must be met:

R_1 The explanans must be true.
R_2 All the statements of the explanans must be verifiable.
R_3 The explanans must contain a law.
R_4 The explanans must lead to a deduction of the explanandum.

The reason why scientific theories lead to predictions and explanations, then, is their ability to enable the scientist to *reason* about the event in question. This reasoning consists of deducing that the event would occur given earlier conditions and the laws of nature. Putting it another way, understanding a phenomenon is the ability to subsume it under a law(s).[4] An explanation of an event, e, would look like this:

1. $C_1 \ldots C_n$ (A set of initial conditions) ⎫
2. $L_1 \ldots L_n$ (A set of laws of nature)   ⎬ *Explanans*
                                                     ⎭

∴ e (The explanandum event)   *Explanandum*

For example, suppose we wanted to explain why a rock is moving toward the ground at a velocity, v, at time t. Prior to t, the rock was released in a gravitational field (C_1); it was thus subjected to a net force (Newton's law of gravity, L_1); by Newton's second law (L_2), it would accelerate depending on its mass (C_2); given other conditions, and how long the rock was subjected to the gravitational force, we can deduce its velocity at t. This is why the rock was moving at velocity v at time t. Of course, explanations would vary in (deductive) completeness, depending on the extent we can fill in $C_1 \ldots C_n$ and $L_1 \ldots L_n$.

Note that we can use laws of nature to explain *other* laws. For example, we can use Newton's second law and the approximation that the gravitational field is constant to explain (= deduce) the law of freefall, $S = 1/2gt^2 + V_0 t$ (where S, g, t, V_0 denote the distance travelled in freefall, the gravitational constant, the time of freefall, and the initial velocity of the falling object).

Much hinges on R_3, for laws are those essential features of a scientific theory that enable it to explain and predict. Without them, we could not logically connect initial conditions to the explanandum. We said earlier that Hempel and the formalists are committed to using logic to capture the language of science. What is the logical syntax of a law statement? According to Hempel, laws are of a universal form and they are usually conditional in nature. Universal conditionals are propositions of the form (For any X, if X has the property F, then X has G), or 'Whenever X has F, it has G' (technically, $(x)(Fx \supset Gx)$). So, for example, 'All swans are white' is of the universal conditional form and is thus a nominee for being a law of nature. Newton's second law would read: for any body X, if X has a mass, m, and X has a net force, F, acting upon it, then X will accelerate, $a \supset F\ m$. Again, in the formalistic spirit, whether a proposition expresses a law of nature will depend on its logical syntax alone, and not on the context in which it occurs.

In the case of explanations that occur in scientific systems that are probabilistic or stochastic, instead of explaining by showing that the phenomenon in question *had* to occur given the initial conditions and the laws of nature, we explain by showing that given the initial conditions and the laws of nature in a probabilistic form, the event was *very likely* to occur. So, for example, we might explain why an individual contracted a disease by noting that it was very likely that said individual would get the disease in light of the fact that he was exposed to several people who had it. So, probabilistic explanations take on the form:

$P(C_i, e_j) = 0.9$ (The probability that e_j would occur under conditions C_i is very high)

C_i (C_i occurred)

∴ probably e_j (e_j is very likely).

Hempel has argued that his type of explanation is to be distinguished from the D–N kind because the reasoning from C_i to e_j takes place not by means of deduction but by induction. The more positive instances we can find in support of e_j occurring under conditions C_i, we are inclined to say that e_j is more likely to occur under C_i. Thus, we have two types of explanation, each corresponding to a kind of inference.

This is really the key to the formalist approach to theories. *Scientific explanation is really a kind of inference*, a claim many philosophers of science dispute. Scientific understanding, then, not only requires knowledge of the facts but the ability logically to connect them to others by means of laws. But is not this precisely the same kind of understanding that goes on when a scientist predicts that an event will occur? After all, what is a scientific prediction? We start out with a set of initial conditions, $C_1 \ldots C_n$, along with laws, $L_1 \ldots L_n$, and deduce the truth of e at some future time. But, then, the logic of prediction does not differ at all from that of explanation. In fact, the only difference between explanation and prediction is that in the former case, the event has already occurred while in the latter, the event has yet to occur. This is Hempel's famous and controversial *structural identity hypothesis*.[5] It is a direct result of his D–N model, and it has two significant sub-hypotheses: if e can be explained, then e could have been predicted; if e can be predicted, then e can be explained once it occurs. What this means is that if Hempel is correct about scientific theories, then if a scientist can predict a specific range of phenomena, he has, by that very fact, explained them. And many scientists feel that this is absolutely correct: if predictions are generated and confirmed, explanations take care of themselves. So, why even worry about explaining, they say, if we can predict? Later on, I shall present reasons why they should worry because, I argue, contrary to the structural identity hypothesis, prediction and explanation do *not* have the same logic.

2. IMPLICATIONS OF THE D–N MODEL

If Hempel's analysis is correct, then it appears to be the case that many kinds of explanations that occur in everyday contexts, in history, biology and in the social and behavioural sciences are either pseudo or incomplete. Consider the variety of explanations we come across in history or even explaining events that take place in our ordinary lives. They are usually presented in the *narrative* form: *Q* occurred because *P*. For example, 'The automobile crashed because the brakes failed and it could not stop before hitting the wall', or 'The proposition passed because the electorate could no longer tolerate high taxes.' Notice how laws are conspicuously absent in these narrative accounts; they equally appear to lack the predictive power required by the structural identity hypothesis, for even though we can explain the event in question once it occurs, predicting its occurrence on the basis provided by the explanans alone is quite another matter. We very often present explanations of the '*Q* because *P*' form without having these laws at our disposal. What are the laws of history, politics, etc.? They are difficult to find and, according to D–N, these everyday accounts are rendered as incomplete explanation 'sketches'.[6] Their completion requires adding initial conditions and laws to *P* in '*Q* because *P*' until it can be shown that *Q* logically follows from *P* (and could have been predicted).

The same holds for so-called teleological explanations, i.e., explanations of the form '*x* behaved *y in order* to get *z*'. Again, if we explain why 'Jane cut her class' by saying 'because she wanted to spend that time to complete an essay that was due the next day', where are the laws connecting explanans and explanandum? Another example occurs in biology: the heart beats *in order* to circulate blood throughout the body. (These are sometimes referred to as functional explanations.) Again, laws are absent. Now Hempel has a way to handle these cases. *What* exactly is being explained here? If it is why the blood circulates, we can easily cite biological initial conditions and laws which readily fit D–N. But the above functional account purports to explain why the heart beats, and it does this by citing some biological function, end or purpose which the heart serves. Unless we believe in purposes in nature, that ends somehow determine how things work, such accounts are not to be taken seriously, except,

perhaps, as abbreviations for more complicated lawlike explanations.[7] In other words, any so-called teleological explanation of
the form 'x did y in order to get z' can be translated or reduced
roughly:

1. x wants z.
2. Anyone who wants z, under circumstances $C \ldots C_n$, will do y.
3. $C_1 \ldots C_n$ occurred.

$\therefore x$ behaves y.

This is what is really meant when we say 'Jane cut her class in
order to complete her essay.' We are just too 'lazy' to fill in all the
rest. If so, teleology or explaining by appealing to ends or
purposes plays no independent role in scientific explanation.

In Chapter 8, I will discuss the nature and logic of the reduction
of one scientific theory to another more comprehensive and
powerful theory. An example of this occurred when the science of
classical thermodynamics, which deals with the behaviour of
heat, temperature, pressure, etc., was reduced by means of the
kinetic theory to statistical mechanics, the science of the statistical
behaviour of large ensembles of Newtonian particles. As a result
of this reduction of one scientific system to another, the macroscopic thermodynamical behaviour of gases, fluids and solids can
be understood in terms of the microscopic behaviour of collections
of atoms, molecules, etc., which obey the laws of mechanics. The
importance of reduction in the sciences can not be overemphasised. For one thing, many believe that this is the only way the
unification of the sciences can come about, and many feel the
incorporation of diverse scientific systems into a single framework
to be one of the most laudatory intellectual achievements of
mankind. Hempel's D–N model supplies us with a very natural
way to view the reduction of one scientific system to another.
Reduction of one system to another amounts to showing that the
laws and results of the reduced scientific system can be derived
from the laws and initial conditions of the reducing science.[8] By
D–N, such a derivation is tantamount to an explanation of the
laws and results of the reduced science in terms of the reducing
science. For example, if psychology were reduced to
neurophysiology, then any law in psychology could be derived
from the laws and assumptions of neurophysiology. This is how

the laws of psychology would be explained in neurophysiological terms. In Chapters 7 and 8, I scrutinise the feasibility of such a reduction.

What are causes and causal explanations according to D–N? Hempel first distinguishes between two kinds of laws, laws of temporal succession and laws of coexistence; only the former type of law can be used to establish causal relations. An example of a law of coexistence would be the relation between the length of a pendulum and its period, while the law of free fall, $S = 1/2gt^2 + V_0t$, is a case of a law of temporal succession. According to Hempel, a cause turns out to be 'some antecedent state of the total system, which provides the "initial condition" for the computation by means of a theory, of the later state that is to be explained'. So, the cause of the rock travelling a distance S in freefall is the *fact* that it was released with an initial velocity V_0 so many seconds prior to its arrival S units from its point of release. These initial conditions and the law of freefall lead to a prediction and causal explanation of the rock's journey.[9] On this account, Hempel is in complete agreement with the logical positivists who denied that causes were anything more than events that served as predictive devices. Likewise, he is in accordance with Hume who defined a cause as 'an object, followed by another, and where all the objects similar to the first are followed by objects similar to the second'.[10] Far from being links in a chain of physical connections, causes are simply predictive bases.

Notice that while the D–N model is meant to capture the logic of explanation and prediction in a theory, it makes no claim to capture the logic of a theory's formulation. In fact, many of the formalists believe that there is no logic underlying the development of scientific theories. What goes on in the mind of the scientist in the course of discovery is a matter for the psychologists and historians but not the logicians and philosophers of language. In Chapter 4, we will see that Hanson considers the D–N approach's refusal to supply rules or guides underlying the development theory actually reveals a major flaw in its treatment of theories. For now, I will leave the issue whether or not there exists a logic of discovery until I cover Hanson's approach to theories.

3. THE PROBLEM OF LAWS AND LOGICAL SYNTAX

There are two fundamental assumptions of the D–N analysis of theories: the essence of a theory can be captured by its logical syntax; and the essence of explanation is the subsumption of the phenomenon under a law(s) of nature. If both assumptions are to hold, then logical syntax should capture those essential features of lawlike propositions as well. As I pointed out in the last section, a formal analysis of a law sentence will most often be of the universal conditional form, (x) $(Fx \supset Gx)$. Let us first examine the conditional part of these propositions.

The logical symbol often used to represent propositions of the conditional form is the 'horseshoe' or $\supset$. Let P and Q stand for propositions or sentences. If we want to say that 'P implies Q', 'If P then Q', 'Whenever P is true so is Q', 'The truth of P is sufficient for the truth of Q', and so on: each of these can be expressed, '$P \supset Q$', where P is the antecedent and Q, the consequent. '$\supset$' is known as a truth functional connective, i.e., the truth of $P \supset Q$ depends entirely on the individual truth values of P and Q. Suppose we wanted to falsify $P \supset Q$. This would amount to showing that P is true while Q is false. This combination of individual truth values negatively defines $P \supset Q$. Since we are assuming a two-valued logic, i.e., any proposition is either true or false, we must assign a truth value to $P \supset Q$ for all other truth functional combinations of P and Q; and since it appears that the only way to falsify $P \supset Q$ is when P is true and Q false, all other truth functional combinations of P and Q yield true for $P \supset Q$. This, then, is how '$\supset$' is defined in logic. The following truth table represents such a truth functional characterisation of conditional sentences:

P	Q	$P \supset Q$
T	T	T
T	F	F
F	T	T
F	F	T.

'$\supset$' has also been called material implication, and it is what we must use to capture the logical syntax of lawlike statements.

One major difficulty with using symbolic logic to capture the essence of lawlike statements concerns the fact that a false antecedent or a true consequent, alone, suffices to establish the truth of $P \supset Q$ according to the above truth functional characterisation of 'P implies Q'. But the truth of a law of nature can not be established simply by showing that its antecedent is always false or its consequent always true. For example, 'If baseballs travel faster than the speed of light, then they will glow bright orange' and 'If there are martians, then they will not travel faster than the speed of light' would each automatically become lawlike sentences simply because it is true that nothing travels faster than the speed of light. Cases like these are known as *paradoxes of material implication*.

In relation to this, there exists a class of scientific propositions known as counterfactual conditionals or contrary-to-fact conditionals. Their form is 'If X *were* the case then Y *would* be the case', 'Even though X did not occur, were it to have occurred, Y would have occurred', 'If there were an X, then there would be a Y', and so on. Counterfactual conditional statements, then, are not about actual states of affairs but instead about what *would* have been the case even if the antecedent conditions were not, in fact, realised. Scientific theories can always be made to yield counterfactual claims. For example, suppose that there are no bodies in the universe which are completely free of any and all forces of nature. Nevertheless, Newton's first law states how such bodies will behave. In biology there may be certain genetic combinations that have yet to occur in nature; yet, someday, we may have laws of genetics that will tell us what such combinations would look like. Not only does this type of a claim crop up everywhere in science, it appears to be inextricably linked with lawlike sentences. But these statements that are so crucial to science are highly resistant to the truth functional analysis of logicians. For example, if we use material implication to capture the logic of counterfactual conditionals then, in the case where no bodies are free of forces, 'If a body were free of forces, it would move at a constant velocity' and 'If a body were free of forces, it would accelerate' are *equally true*. But our theories somehow tell us that they can not both be true. For decades, philosophers have tried to come up with a truth functional analysis of these statements with, I contend, little success. If '$\supset$' can not capture *their* meaning, what can we conclude about logical syntax being the essence of theory?

Turning now to the 'universal' part of the universal conditional analysis of lawlike sentences, Hempel and the formalists are quite aware of another problem confronting them: there are many propositions of the universal conditional form that are clearly not lawlike but are 'accidental' universal conditionals. 'All the nails in my pocket are rusty', 'Whenever the 5:00 p.m. train passes the factory, the workers leave', and 'High amounts of fluoride in drinking water are correlated with high abortion rates', may be true statements of the universal conditional form but they certainly do not qualify as laws of nature. The problem for the formalists is to differentiate genuine lawlike sentences from accidental universal conditionals on the basis of their logical syntax. This is not a very easy thing to do. Suppose we say that laws are universal conditionals that are unrestricted in universality. So, for the rusty nails in the pocket case, if we restate the conditional as 'Any object made of iron placed under conditions $C_1 \ldots C_n$ will oxidise' then we may just have a law of nature here. But opponents will argue that we would only consider it to be so because of our knowledge of chemical theory, i.e., not any universal conditional of unrestricted generality is a law but only those that are supported by theory. If so, we can not analyse a theory in terms of a set of laws which, in turn, are analysed as universal conditionals.[11] In fact, they maintain, it is the other way around: we must use theories to identify the laws, to sort out the genuine universal conditionals from the accidental ones.

4. THE THEORETICIAN'S DILEMMA

In his formulation and discussion of the theoretician's dilemma,[12] Hempel clearly parts with the logical positivists. The latter maintained that, contrary to the realistic conception of theories, theories did not entail anything ontological; theoretical terms such as 'atom', 'electron', 'the ego', etc., do not really denote anything real about the world but are, instead, 'convenient myths' used solely to organise experimental data. This stance on the ontological status of theoretical terms can be nicely expressed in the form of a dilemma:

1. Either theoretical terms serve their purpose or they do not.
2. If they serve their purpose, they are unnecessary.

3. If they do not serve their purpose, they are surely unnecessary.
4. ∴ Theoretical terms are unnecessary.

The further conclusion is that if theoretical terms are unnecessary, we have no scientific reason to believe that the theoretical entities referred to by these terms exist. They are to be discarded as 'excess baggage' since theories can function, i.e., predict and explain observational results, without having to postulate their existence.

Of course, the most controversial premiss in the above argument is the second one. Let us see how the logical positivist would argue for it. Using the weak version of the verification principle, even if we can not observe a theoretical entity, we must be able to state empirical conditions that suffice to tell us if and when the said entity exists; likewise, the existence of our theoretical entity must have observable effects. Letting 'O_i' and 'O_f' respectively stand for initial and final observable conditions while 'T' stands for the occurrence of a theoretical entity, the above requirements look so:

1. $O_i \supset T$
2. $T \supset O_f$

But 1 and 2 logically entail $O_i \supset O_f$. In other words, if theoretical terms are to play a role in the organisation of scientific data, they are unnecessary, for $O_i \supset O_f$ does the same (deductive) job without them.

Hempel's reply to the above argument is twofold. In the first place, contrary to what 1 and 2 supposedly imply, the meanings and conditions for the application of theoretical terms can not be exhausted by observables. This point is actually connected with the paradoxes of material implication. Recall how the logical positivist proposed to reduce the meaning of theoretical terms to those of observables: the meaning of a theoretical term denotes a matrix of test conditions and results, i.e., 'T' means $(C_1 \supset R_1)$ and $(C_2 \supset R_2)$ and $(C_3 \supset R_3)$ and ... or 'T' $\equiv (C_1 \supset R_1)$ and $(C_2 \supset R_2)$ and $(C_3 \supset R_3)$ and .... The above reads that the theoretical property occurs if and only if we get a positive result for each and every test of its occurrence. To say someone is intelligent, for example, means that if you give that person an IQ test, he will score high; if he is placed in a problem-solving situation, he will solve the problem quite rapidly in comparison to others in the same situation; if placed in a completely now situation, he will adapt to it at a faster rate than the average individual; and so on. However, suppose we do not place the individual in *any* of the $C_1 \ldots C_n$ test situations. Then each material conditional on the right-hand side is true and, on the basis of that alone, we must ascribe the theoretical property of high intelligence to the individual! Surely, we do not want to ascribe theoretical properties simply because the individual systems are not tested for them. Now Carnap and Hempel were well aware of this problem, and they concluded that a complete definition of theoretical predicates in terms of observation or test sentences can not be given; at most, we have only a *partial* characterisation of them in terms of test sentences if we are to avoid the above paradoxical ascriptions of properties to things.[13]

The moral of the above story is that the meaning of theoretical terms should be left 'open' or independent *vis-à-vis* observables or test sentences. The goal of the operationalists will never be achieved because we can not specify necessary and sufficient conditions for the application of theoretical terms by using test sentences without falling prey to paradoxes of material implications. (There are other considerations, such as the results of Ramsey and Craig, but they need not be discussed here.) Hempel's second point against the theoretician's dilemma is that it falsely assumes that the sole purpose of a theoretical term or a theory is to organise observables by establishing deductions among test sentences. The above argument neglects the fact that theoretical terms, indeed, serve *other* purposes:

> When a scientist introduces theoretical entities such as electric currents, magnetic fields, chemical valences, or subconscious mechanisms, he intends them to serve as explanatory factors which have an existence independent of the observable symptoms by which they manifest themselves.[14]

In opposition to the logical positivists, we need ontology or the postulation of entities which are not directly observable in order to explain observation results. Thus, Hempel has answered the theoretician's dilemma.

Or has he really answered it? Let us scrutinise his reply in more detail. In the first place, the above critique against the operationalist reduction of theoretical terms is actually a red herring. The operationalist runs into the above counterexamples only if he chooses to use material implication or symbolic logic to express his operational characterisation of theoretical terms. But operationalists such as Mach, Bridgman or Hull (in psychology) need not use the '$\supset$' to define; they can do it simply by using the calculus already provided by the scientific framework in which they are working. Why should their enterprise be wedded to translating the language of science in terms of symbolic logic? The paradoxes of material implication are more an indication that truth functional logical syntax is not a good device for analysing the language of theory than showing that operationalism is doomed to failure.

Even more damaging, Hempel's above reply rests on a mistake. He is assuming all along that 1 and 2 entail some form of operationalism; as a result, most of his efforts are devoted to showing how one can not reduce the meaning of theoretical terms to test sentences among observables. But 1 and 2 do not require such a reduction. In order to show this, we must first make an important distinction. It is one thing to say that theoretical terms must have empirical necessary and sufficient conditions for their application; this is precisely what 1 and 2 say. We can call 1 and 2 an *empiricist* requirement for the use of theoretical terms. But it is quite another thing to say that the application of a theoretical term has necessary and sufficient conditions which consist of conditionals connecting observables *only*. This is a stronger, *positivistic* requirement placed upon the application of theoretical terms, one which follows from the verification principle of meaning. (Logically, it is the difference between [$(T \supset O_f)$ and $(O_i \supset T)$] and [T if and only if $(O_i \supset O_f)$].) Since the conjunction of 1 and 2 in the theoretician's dilemma does not entail the second, positivistic requirement, refuting the latter does not lead to a refutation of the former.

In light of this mistake, not only does Hempel's first criticism break down, his second reply places his own D–N model in deep waters. The second answer essentially attempts to show how theoretical entities and their corresponding terms play an indispensable role in explanation. But the D–N model of explanation says that the logical syntax of a theory, its laws in particular, is responsible for its explanatory power. Where, then, does the ontology of 'theoretical entities such as electric currents, magnetic fields, chemical valences, or subconscious mechañlsmo' fit into the explanatory picture? It seems that Hempel's only recourse is to say that postulating the existence of a theoretical entity enables the scientist to explain facts or data that could not otherwise be explained by observables alone. However, by his D–N model of explanation, this translates into: there are certain factual or observable sentences that could not be inferred from laws and initial conditions unless there were theoretical terms among these laws and initial conditions. In its simplest form, it says that there is an observable, O, that could not be inferred within a scientific framework unless there exists a law statement(s) with the occurrence of a theoretical term, T, in its antecedent:

1′. $T \supset O_f$ (law connecting a theoretical entity with the observable)

2. T

∴ 3. O_f (without 1′ and 2, O could not be inferred)

This appears to be the logic of saying that theoretical entities are needed to explain if we are to keep within the confines of D–N.

Here lies the rub in the above answer. Even if we reject the positivist requirement, the empiricist requirement for the application of a theoretical term may be a reasonable one for us to accept. After all, what science does not state empirical conditions, deterministic or probabilistic, for the existence or occurrence of its theoretical entities? These statements express links between prior, observable conditions and the subsequent occurrence of a theoretical entity. For example, a psychologist might claim that a certain, observable physical event will cause specific non-observable mental events. In fact, stating connections such as these is one, very important role of scientific theory. If so, we should always have a proposition of the form, 2′$O_i \supset T$ (where 'O_i' denotes an initial observable) and, once more, when we add Hempel's claim that theoretical terms are necessary for explanation, or 1′, T drops out as the middle term of a hypothetical syllogism. Since 1 and 2, 1′ and 2′ are logically equivalent, we are back at square one. If explanation is a matter of inference, why postulate theoretical entities if we can infer without them? So, if D–N is correct, one wonders how theories can be anything more than epistemological systems that do nothing but systematically handle data (as the logical positivists maintained all along). Not only do I consider

Hempel's reply to the theoretician's dilemma lacking: his own account of the nature of scientific explanation appears to undercut the need for theoretical entities.

In the light of the above failure of D–N to answer the theoretician's dilemma, I feel that in order fully to appreciate the role ontology plays in science and the indispensability of corresponding theoretical terms – in particular, how theoretical entities lead to an understanding of nature – we must free ourselves from the picture that a scientific theory is merely an epistemological system or an algorithm for the organisation of data. In the ensuing chapters, I show how the ontology of a theory *vis-à-vis* its logical syntax yields powerful and sometimes even beautiful explanatory insights into nature. Remember: for Hempel, laws are the essence of explanation, and they lack ontological significance because they do not refer to causal connections in nature. However, if my view that logical syntax does not capture the essence of explanation is correct, then it should be possible to construct legitimate and complete explanations where laws are conspicuously absent. This possibility will now be explored.

5. EXPLANATIONS WITHOUT LAWS

Dray[15] and Scriven[17] were among the first to argue against the fundamental assumption of D–N that laws are essential to explanation. Dray argues that there are countless explanations in history which are of the narrative form, where not only is it clear that laws are absent in the explanans, it is highly dubious that one could come up with a law in history such that when it is added to the narrative account, it would lead to a more relevant explanation. Scriven's case against laws being the essence of explanation is based upon a distinction he makes between the *content* of an explanans and the *grounds* one uses to support its truth. If Scriven is correct, then laws are not the essence of explanation but what we use to support explanations – whether these explanations occur in everyday situations, history or even physics.

Scriven presents us with an example of a narrative explanation which does not contain a law: the ink bottle is on its side because John knocked it over. He claims that even if we were ignorant of any laws of physics which could be used to support such an explanation, we still understand why the ink bottle is on its side.

The explanans is complete without them, and to demand more is simply to confuse the content of the explanans with support for its truth. Taking up Scriven's point, let us make a list of explanations of the narrative or '*Q* because *P*' form:

(a) The ink bottle is on its side because John knocked it over.
(b) The piston moves back and forth because the connecting rod pushes and pulls the piston back and forth.
(c) *A* and *B* stay together because the bar holds them together.
(d) Protons and neutrons remain in the nucleus because π-mesons bind them in the nucleus.[17]
(e) The rods move at an equal angular velocity because the gears maintain their rates equal.

Scriven claims that statements like (a)–(e) constitute counter-examples to Hempel's model.

Hempel's defence is based on the notion of explanatory *relevance*. Not *any* fact statement '*p*' can be used to explain an event statement '*q*'. For example, the fact that the moon has craters is not used to explain why the inkwell is lying on its side. On the contrary, we cannot know or even express that '*p*' is *relevant* to explain '*q*' unless it is logically connected to '*q*' by means of a law. So, if laws are essential to establishing the explanatory relevance of a '*p*'-statement, then Scriven's distinction between the grounds of an explanation and its content breaks down. General laws not only serve to justify the connection between '*p*' and '*q*', they are also what makes the '*p*'-statement explanatory.

Hempel thus supplies Scriven with a challenge: if D–N is inadequate because of this corpus of so-called explanations of the form '*q* because *p*' then the onus is on Scriven to supply a competing model of explanation, one which will present us with a rival conception of explanatory relevance. Otherwise, Hempel can level the charge that '*q* because *p*' explanations are only apparent counterexamples to D–N because they are presented in such a vague fashion; they require much more analysis if they are to constitute a genuine class of counterexamples. After presenting this challenge, Hempel confidently concludes that any analysis of '*q* because *p*' type explanations will no doubt end up with his model, nomological statements and all.[18]

Can we take up Hempel's challenge? In order to do so, we must first show that '*q* because *p*' statements [such as (a)–(e)] are not vague as Hempel claims. Secondly, a proper analysis – one other than D–N – of '*q* because *p*' statements must be supplied; and finally, it must be shown that the explanatory relevance of a '*p*'-statement is independent of its being connected to '*q*' by means of a law. These tasks will be taken up now and in the next section.

To begin, one thing to notice right away about (a)–(e) is that each sentence has a definite syntax. For example, each '*p*' statement possesses a transitive verb such as 'knocks', 'pushes', 'pulls', etc. This seems to be a rough version of the syntax of '*p*'-statements in a '*q* because of *p*':

Explanans ('*p*'-statement): Noun Phrase + Transitive Verb + Direct Object + Objective Complement.

For example, 'John knocked the inkwell on its side' is of this form. The objective complement in the above formula has a very important role. I am claiming that '*p*'-statements explain by describing events in terms of a causal process that takes place between the objects denoted by the noun phrase and direct object term while the transitive verb describes the nature of the causal connection. The transitive verbs in the above equation are called 'causative' verbs. Not all expressions that contain transitive verbs are causative, however. For example, transitive verbs such as 'know', 'found', 'call', 'thinks', etc., are not causative and do not denote causal processes. The distinction between causative and non-causative transitive verbs can be made by noting that the former take objective complements while the latter do not. (Objective complements complete the description of the action expressed in the verb and modify or qualify the direct object.) I maintain that if a '*p*'-statement is meaningful and contains a transitive verb and an objective complement, then that statement in itself can serve to explain the '*q*'-statement connected with it – without having to appeal to a law sentence. This is one way to explain.

So, in these cases, the '*q*'-statement is connected to the '*p*'-statement of the narrative account, not by means of a law, as Hempel maintains, but because of the fact that '*p*'-statements have an unique internal structure.[19] The syntactical features of '*p*'-statements are such that they generate or entail their corresponding '*q*'-statements in accordance with this transformational formula:

> Explanandum ('*q*'-statement): Noun Phrase (= Direct Object of '*p*' statement) + Predicate (= Copula + Objective Complement of '*p*'-statement).

By 'copula' is meant a linking verb that expresses a state of being and connects its subject to the predicate. The copula appearing in the '*q*'-statement will depend on linguistic content. For example, if we start with the '*p*'-statement in (b) and use the linking verb 'moves', we can see how '*q*' ('The piston moves back and forth') can be generated from '*p*' ('The connecting rod pushes and pulls the piston back and forth'). In (c), the appropriate copula would be 'stay' for we wish to explain why A and B stay together as opposed to separating. So, from 'The bar holds A and B together', we can generate 'A and B stay [are] together'.

Why am I going to all this linguistic trouble about '*q* because *p*' explanations? For one thing, I have satisfied the demands of the formalists to give definite form or structure to narrative explanations. To some extent, doing so mollifies Hempel's charge that Scriven's counterexamples are vague, for they do have a definite logical syntax, even if it is not that of D–N. But even more important, we now have a syntactic criterion of explanatory relevance, one which is not parasitic on laws – something Hempel said could not be done. The relevance of a particular '*p*'-statement for explaining a '*q*'-statement depends whether '*p*' and '*q*' are transformationally related in accordance with the above formula. The reason why the existence of craters on the moon is not explanatory *vis-à-vis* the inkwell being on its side is because we cannot supply a suitable causative transitive verb between the corresponding noun phrase and direct object.[20] The above transformational relation between '*p*' and '*q*'-statements is thus the rival model that Hempel called for when he attacked the notion of a narrative type of explanation. This is just a syntactic model, however, one that is primarily intended to generate counterexamples against the hypothesis that explanations explain because they contain laws. A more general and positive approach to explanation is presented in Chapter 7.

The above transitive verb model of narrative accounts is not intended, then, to be a characterisation of causal explanations.

But it does give us an interesting clue as to the nature of causal relations, one that differs significantly from Hempel's characterisation of a cause as an initial condition which, along with laws of succession, leads to the prediction of the effect.

6. CAUSAL DETERMINATION

Thus far, we have seen that the major contention of the D–N approach is that explanations are a kind of inference; in particular, causal explanations are inferences from past and present conditions to future events, conclusions which are based on laws of temporal succession. This renders causes as potential predictions. I have been questioning this assumption all along, and in this section, I go against the Humean tradition by arguing that causal explanations are not inferences but statements that refer to processes in nature which are 'stronger' than the mere correlations of events. If I am correct about this, it means that scientists postulate causal relations just like any other theoretical entity such as electrons and photons. Causal connections and theoretical entities such as the particles of physics have the same ontological status. But this means that causal relations are more than just the joint occurrence of events. Contrary to the Humean view, it is not enough to show that one event followed another in order to establish that a causal relation has taken place; more must be shown, even if it can be demonstrated in addition that these two types of events are correlated. Below I will present a model of causality which is intended to make sense of a relationship which is stronger than correlation.[21]

The question I am seeking to answer here is whether a complete analysis of causation can be accomplished without bringing in an element which goes beyond the mere correlation of events.[22] Before I answer this question, however, let us consider a well-known counterexample to Hempel's analysis of causal explanation. It concerns the direction of the causal relation. Recall that Hempel adopts the Humean view of causation. By that we mean that the cause is an event which regularly occurs before the effect event. According to this view, the direction of the causal relationship is simply determined by the temporal ordering of events. Now everyone agrees that the causal relation is irreflexive (*a* does not cause itself) and asymmetrical (If *a* causes *b* then *b* does not cause *a*). But so is '*a* came before *b* in time'; so it is quite natural to use this relationship to capture these logical features of causal propositions. This is known as the regular succession view of causality, and because the temporal relation of *a* (the cause) *came before b* (the effect) captures the essence of causal determination, it follows that causation does not occur in a single instance.

In his article 'Causation and Recipes',[23] Douglas Gasking attacks the Hume–Hempel regular succession view.

> But this 'regular succession' notion will not do. For there are cases where we would speak of A causing B where it is not the case that from the occurrence of A we may infer the subsequent occurrence of B.

This is because the cases Gasking has in mind are those in which 'the effect is not something subsequent to the cause, but simultaneous'.

> From the fact that a bar of iron is now glowing we can certainly infer (and it will be as causal inference) that it is now at a temperature of 1,000°C or over. Yet we should not say that its temperature was caused by the glowing: we say that the high temperature causes the glowing, not vice versa.

There are numerous other cases exactly like the above. Take Newton's second law, for example, which is clearly a causal law connecting forces and the accelerations they bring about; but this law is time independent and, as any physicist knows, this means that cause and effect are simultaneous. Thus, in cases such as these, we can no longer use the 'before' relation and laws of temporal succession to determine the direction of causation, i.e., capture the above-mentioned asymmetrical and irreflexive aspects of the causal relationship.

The above problem of determining the direction of causality poses what I understand to be insurmountable difficulties for the D–N account of causal explanations. First of all, the above cases will not go as long as causal terms occur in laws that are time independent. Such laws occur throughout the physical sciences. (It will not do at all to 'make up' delaying factors in, say, the second law, for doing so would lead to a violation of the conservation of momentum.) One may even wish to deny a causal status to forces but if forces do not qualify as a paradigm of causation, what does? Even more important, in these cases, causal explanations no longer fit the D–N inference paradigm precisely because the laws which contain causal terms are both time independent and symmetrical. This means that while we expect causal explanations to be asymmetrical, D–N can not capture this aspect of them, for in the above cases, we can freely go from *a* to *b* or from *b* to *a*. Thus, suppose we observed two rotating gears, *a* and *b*:

Are we to say that *a*'s clockwise motion caused *b* to rotate counterclockwise or is it the other way around? An answer to this question in terms of what we can infer about the motion of *a* on the basis of *b*'s motion or vice versa is not forthcoming because *a* turns clockwise if and only if *b* turns counterclockwise.

If neither the direction of inference nor temporal order determines the direction of causality in these cases of simultaneous causation, then what does? My answer is that *it depends on the ontology of the situation*. In order to show this, I will present a model for the causal relationship, one that makes sense of its asymmetrical features without having to rely on temporal succession.

First of all, it appears that there are contextual features surrounding the concept of 'cause' which ought to be drawn out if we are to appreciate the ontological features of the causal relationship. I think that it is an unfortunate outcome of the philosophical dictum, 'Every event has a cause', that any kind of change turns out to be a caused change. This does not seem to be the case in the theoretical sciences, for they clearly distinguish changes in objects that take place independently of other objects with changes that result from interactions with other objects; only the latter type of change is a causal one. In other words, there are certain changes that take place in individuals that can be completely accounted for without having to appeal to the behaviour of other individuals, and I will call these changes 'natural changes'. Theoretically, an object may undergo a change that is free from or independent of external constraint. This is another way of saying that the cause is external to the change it brings about.

In order to pursue this further, let us examine some examples of natural change. In Feynman's *Lectures on Physics*,[24] the law of inertia is stated in this way:

> If an object is left alone, is not disturbed, it continues to move with a constant velocity in a straight line if it was originally moving, or it continues to stand still if it was just standing still.

The key phrase here is 'left alone'; bodies take on a unique behaviour pattern when left to themselves. Now, in the same way that Galileo regarded circular motion about the centre of the Earth to be 'entirely natural and self-explanatory', Newton would reject a request for a causal explanation of constant linear velocity, and Einstein would not search for the cause of motion along a geodesic. So a caused change is in contrast to a natural one.

> In '*a* causes *b*', '*b*' designates a change in an object, a change which is an *unnatural* one.

This means, however, that causal explanations are relativised to a theory, for what may be a natural change in one system will turn out to be an intervention in the natural course of events in another.

With this qualification of the use of 'cause' in hand, I come to the ontological aspects of the causal relation. Throughout the history of science, there are fundamental quantities that have been introduced by the scientist which enjoy the feature of being conserved throughout physical interactions. For instance, in classical mechanics, momentum and energy are conserved. Once these qualities are ascribed to objects an 'accounting' problem comes into existence when bodies interact. For how are we to keep track of them before, during and after a physical interaction takes place? For the purpose of illustrating this problem, consider the following thought experiment. It consists simply of two spheres, *a* and *b*, one moving and the other at rest, which undergo a sequence of changes with regard to their respective momenta. Sphere *a* moves up to *b* which is at rest, makes contact with *b*, and then *b* moves off while *a* ends up at rest. In other words, *a* lost all the momentum it once had while *b* gained momentum it did not have before contact with *a*. How did *b* get its momentum at time 3?

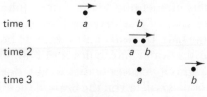

time 1   *a*   *b*

time 2   *a b*

time 3   *a*   *b*

FIGURE 3.1

The obvious answer is that it was transferred from *a*. In other words, a quantity that *a* had at one time *b* possesses at a later time. This makes *a* a cause and *b*'s motion the effect. So we have this ontological feature of the causal relationship:

> Prior to the time of occurrence of the effect, the cause object possesses a *quantity* (e.g. momentum, kinetic energy, heat, etc.) which is transferred to the effect object and manifested as the effect.

In the above case, the transference of momentum from *a* is manifested by *b*'s motion at time 3.

Return now to the problem of causal directionality. Above, I asked: if neither the direction of inference nor that of the *before* relationship determines the direction of cause and effect, then what does? The answer supplied by the above condition is that the direction of transference does; it depends on what object gives up a quantity (the cause) and what ends up with it (the effect). In the case of the gears, if we believed that gear *a* transferred its angular momentum to *b* and not vice versa, we would say *a* caused *b* to turn. The reason why we say heat causes iron to glow but not the other way around is because while we have theories that tell us of the existence of processes which convert heat energy into the molecular energy of the iron which, in turn, is converted into light energy, we know that the reverse of these processes of the transference of energy is highly unlikely. (Imagine light coming out of nowhere, i.e., having no known source, all of a sudden converging upon the piece of iron and heating it up to an extremely high temperature. A likely story, indeed!)

There are countless applications of this transference model in the sciences and for everyday situations. 'John knocked over the inkwell' has us picture John's hand imparting a rotational motion to the inkwell. The cue ball causes another billiard ball to move in that it transfers its momentum to the latter. 'John threw the ball away' is another case of transference of motion, this time from John's hand to the ball. In 'heat causes water to boil', heat energy is transferred to the water molecules and manifested as kinetic energy. Finally, 'gravity causes bodies to fall' does not *mean* that we predict that bodies will fall on the basis of their being placed in a gravitational field. 'Gravity causes bodies to fall' has us envisage this process: the sun transfers its momentum to a surrounding gravitational field which, in turn, transfers its momentum to bodies; and this transference is manifested by their falling.[25]

Let us return to the problem of keeping track of quantities throughout the interactions of bodies. One puzzle to consider concerns the question of the *identity* of these quantities. To say they are transferred suggests that the quantity a had at time 1 is numerically the same quantity b had at time 3. If we were to deny this, we would have quite a mystery on our hands, for then we would have to say that somehow a lost all of its momentum at time 2 and somehow b acquired an equal amount of momentum at the same time! It would be much more natural to regard the loss of momentum in a and the gain in b as stages of a single process of transference of one and the same quantity rather than as separate and distinct events.[26] Besides, we have yet another advantage in maintaining the identity of quantities, here, for not only do we explain why b moved, but we have also explained why b moved when and where it did. In other words, b moved at time 2 because, at time 2, a's momentum was transferred to b.

If the above model of causation is correct, then we now have an answer to the question posed at the beginning of this section: is there an element to the causal relation in addition to mere sequence of events? The answer is 'yes'. The causal relation is more than sequence in that the cause and effect are not only objects or events which are constantly conjoined but, *in a single instance*, the cause object transfers something to the effect. Since this process involves numerical identity of that particular quantity throughout the sequence of events, it can be seen how, ontologically, causation is stronger than correlation. This also means that causes are more than potential predictions.

What are causal explanations, then, if not inferences based on laws of temporal succession? They are narrative accounts of change that describe the transference of a quantity from the cause object to the effect. Although, in the course of such accounts, we may support our causal claim by appealing to a law of nature, such general statements are not part of the meaning of what took place ontologically during the interaction. If laws are not part of the meaning of causal explanations, then D–N fails to capture the essence of a class of explanations which are believed by many to be the most important in science, at least, in the physical sciences.

The above view of causation is contrary to the stance held traditionally in philosophy. It has been open to two types of criticism which ought to be covered here. In doing so, it can be seen how the above model fits in quite well with recent developments in physics. (For those who have no familiarity with physics, it might be advisable to go on to the next section.)

The first criticism has been presented quite recently and is part of an overall attack on any notion of causal determination. In 'Causation: A Matter of Life and Death',[27] John Earman presents a series of ingenious arguments which purport to show that any criterion for the direction of the causal relation is at best arbitrary and without foundation. Earman wants to show that because the direction of causality is arbitrary, Hume was correct when he denied the existence of causal efficacy or causal powers. Most of Earman's arguments are directed against using temporal ordering as a basis for determining the direction of causality. With this, the reader knows, I agree. However Earman then turns his attention to the transference criterion of directionality.

At first glance, his criticism appears to be quite penetrating: the transference criterion fails precisely because different observers will claim opposite directions of transfer of the relevant quantity.

> [The transference] story will not always identify the causal agent in an observer-independent fashion, for different observers may see a different direction of transfer of the relevant quantity. (p. 24)

Let us see how this works in the above case of *a* and *b*. In that situation, our observer is at rest with respect to *a* at time 1. Call this the *a*-frame of reference. Now suppose we have another observer whose frame of reference (*b*-frame) is at rest with respect to *b* at time 1. In the *b*-frame, the same sequence of events looks so:

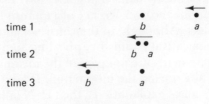

FIGURE 3.2

Thus, the *b*-frame observer concludes that *a* causes *b* to move because of the apparent opposite direction of momentum transfer. Earman appears to be quite right on the above. Note, his criticism assumes that no observer has the right to claim, as a Newtonian would, that either *a* or *b* is *really* at rest at time 1 and the other only apparently in motion. (This assumption is in accordance with the special theory of relativity.) Otherwise, Earman's critique is all too easily answered: either the *a*-frame or the *b*-frame is the right frame of reference but not both. If *a* is really moving at time 1, then *it* is the cause and the *b*-frame description only makes it look like *b* is the cause. But if we are unable to say that either *a* or *b* is really at rest before their interaction, does this then mean that the direction of transference is an arbitrary matter in that what may be a causal contributor in one frame receives momentum and energy relative to another? If so, Earman has supplied a clear-cut refutation of the transference criterion for the direction of causation.

The above criticism appears to be quite strong, for as long as we are denied absolute motion, one *could* always find a frame of reference where *a* is at rest and then find another where *b* is at rest, making it arbitrary which of the two is the cause and which one the effect. What Earman overlooked, however, is that *a body at rest can contribute energy and momentum just as easily as one that moves*; and this is the key to answering his objection to the transference model. In the case where there is no such thing as absolute motion when two bodies collide, *a* does not cause *b* to move alone or vice versa; rather, *a* causes *b* to change its motion *and b* causes *a* to change its motion as well, i.e., each one causes a change in the other, which is manifested by *a* and *b* accelerating. This means that *a* and *b* exchange their momentum and energy. So *two* transferences are involved here. So, my answer to Earman is that causal relations come in pairs when bodies elastically collide. In classical mechanics, it can be easily shown how this momentum exchange is observer independent. Since mass is frame independent while velocity is not, momentum is not invariant. But, the *change* of momentum which occurs to *a* and *b* is. Thus, how much momentum a body gains or loses in an interaction, in classical mechanics, is invariant in all inertial frames of reference, preserving the direction of transference. So, no matter in what inertial frame the observer is located, he will 'see' the interaction the same way as any other observer; they will all conclude that *a* and *b* caused each other to change their motion, independently of whether the causal contributor is moving or at rest relative to a given inertial frame.

Another criticism, one which is also in accordance with the Humean approach to causation, denies that the direction of transference can be used to determine the direction of causation without begging the question. In *Hume and the Problem of Causation*,[28] Beauchamp and Rosenberg maintain that any application of the transference criterion actually presupposes knowledge of causation. In the case of the gears, 'it is no easier to detect a transfer and its direction than it is to determine which of two simultaneous (or non-simultaneous) causally connected events is the cause and which is the effect'.

Unfortunately, the Beauchamp–Rosenberg treatment of the gear case is technically flawed. If they were precise enough about the actual physics of the situation, they could have realised that we can easily determine the direction of transference *independently* of any causal assumptions. We can do this simply by measuring the moments of inertial and respective angular velocities of *a* and *b* before and after the interaction. Assume that before they are meshed *a* is spinning while *b* is not. Suppose *a*'s inertia equals that of *b* and we come upon the gears rotating together. By the conservation of angular momentum, we know that *a* gave up half of its original momentum to *b*, i.e., caused *b* to rotate and converted half of its kinetic energy to heat which shows up as an increase of temperature. All of this can be known simply by keeping track of the motion, inertia and temperature of each gear as a function of time, plus the conservation of momentum. If we could not make these measurements, we could not learn about the direction of transference but it does not mean that transference did not take place or that causal questions are begged here.[29]

While the status of causation may be an open question in

today's most advanced physical systems such as quantum field theory, there are many physicists who use the transference of quantities to make sense of or to model interactions among elementary particles. For centuries before the advent of the new physics, forces were thought to be the paradigm of causation. In Newton's physics, these forces could affect objects over vast distances. But action-at-a-distance is abhorred by contemporary physics, and the transference model is used to avoid it. The way this is done is to replace the old paradigm of causation taking place by one particle exerting a force on another, distant one with an energy- and momentum-laden particle being transferred from the cause to the effect. In fact, all forces in nature are thought to work by the exchange of interacting or 'messenger' particles. For example, the repulsive force which exists between two electrons is reduced in this way. Electron 1 causes electron 2 to be deflected by emitting a photon (the messenger particle) which is absorbed by electron 2.[30] Each force in nature has its own transference particle(s). Once more, causation is more than repetitious temporal succession.

There is another line of defence available to the proponents of the Hume–Hempel view of causation, one that has its roots in operationalism. If the existence of conserved quantities is a source of problems for the regular succession view of causation, we should be able to reduce the meanings of their corresponding theoretical terms to generalisations over observables as Mach did when he reduced the meanings of 'force' and 'mass' in terms of reaction accelerations. Thus causation, once more, can be thought of in terms of generalisations among observables. Not only is the above enterprise of reducing fundamental quantities an extremely ambitious task: I feel that there are good reasons to preserve the independence of meaning of theoretical terms such as 'force', 'energy' and 'momentum'. Returning to Mach's attempt, Max Jammer showed[31] that if we adopt his definition of 'mass', the result is that in a universe consisting of $n \geq 5$ bodies (with three degrees of freedom), it will be impossible to compute the acceleration of each body due to the other bodies in the system. Hence, Mach's system would be incomplete. Likewise, if we restrict the meaning of 'mass' in terms of reaction acceleration ratios, then mass becomes a dimensionless quantity with the result that the unit dimensions of other fundamental quantities in mechanics are systematically changed in a highly unintelligible way. For example, the dimensions of 'force' become $[cm]^2 [sec]^{-2}$ and the units of 'energy' would be $[cm]^{-2} [sec]^{-2}$, which has us identify the energy of a particle with the square of its velocity.

But the major reason I am against reducing the meaning of 'momentum', 'mass', 'energy', and so on is the restrictiveness that results. Einstein's famous relation between the rest mass of a body and its energy, $E = mc^2$ (where 'E' denotes energy, 'm' mass and 'c' the velocity of light), is considered to give us a new and significant insight into the nature of mass and energy. For one thing, it tells us something about these quantities never realised before: besides the fact that a new form of energy – one other than kinetic and potential energy – has been discovered, we now learn that mass can be converted into energy and vice versa. If we accept Mach's definition of mass, however, how can we possibly speak of the conversion of acceleration ratios to energy?[32] It appears that we are forced to say that 'mass' and 'energy' have been redefined by this new relation and that Mach and Einstein must have been talking about different quantities. But then if they end up talking about different quantities, Einstein really did not discover anything new about mass and energy after all, which is quite contrary to what is traditionally taught in physics. The point is that learning something new about a scientific quantity – and this seems to be one indication of scientific progress – presupposes a rough sort of identity between basic concepts. (As we shall see in a later chapter, some philosophers will deny this.) But if Mach's method of reaction accelerations is treated as a definition of 'mass' rather than as just another means of measuring mass – which happens to be my view – any adjustment of a mathematical relation between quantities involving mass becomes false 'by definition'. I think it would be much better, then, to preserve the independence of meaning in these cases in order to allow for future developments of one's scientific system.

Where does this leave us? This rival view of causality has created serious objections to the D–N approach to causal explanation. Notice how this view allows one to claim a causal relation has taken place in a single instance, i.e., without bringing in laws in the content of the causal explanation even though we might appeal to laws to support the explanans. Such a possibility has ramifications for a major implication of the D–N model, namely, the structural identity hypothesis.

7. UNDERSTANDING WITHOUT FORESIGHT

According to D–N, explanation and prediction have exactly the same logic, differing only pragmatically. Hempel divides the structural identity hypothesis into two sub-theses: '(i) *that every adequate explanation is potentially* a prediction ... ; (ii) that conversely *every adequate prediction is potentially an explanation*'.[33] Of these two sub-theses, Hempel appears to have more confidence in (i) while he regards (ii) as more or less an 'open question'. It does appear *prima facie* that (ii) is more readily open to counter-examples than (i). Refutation would be a matter of supplying cases where events are correctly predicted while understanding is clearly absent. For example, even though Kepler's laws enable us to predict the motions of the planets with a high degree of accuracy, one wonders how we could use them to explain planetary motion. One law states that the planets sweep out equal areas at equal times. Thus, we can predict that a planet near the sun will move faster than one further away if it is to sweep out the same amount of area. But how can this law serve to explain *why* said planet moves faster? Other examples of this order can be supplied. A defence would most likely amount to showing either that understanding actually does occur or that a genuine scientific prediction was not achieved.

Counterexamples to (i) are more difficult to come by. Cases supplied by Toulmin, Scriven and others have occurred in the literature. Darwin's theory of evolution may readily explain the establishment of a new species in terms of mutation and natural selection but the biologists could hardly have predicted that a particular new specie would have occurred on the basis of their explanation. This is because they could not have predicted the particular mutation. Likewise, we could explain the collapse of a bridge by appealing to fatigue in a particular beam; however, while this serves as an adequate explanation of the bridge's collapse, it is an *ad hoc* explanation, for 'we simply do not have the data required for a prediction that the failure would take place on a certain date'.

While the above counterexamples to (i) are interesting, I feel, along with Hempel, that they are inadequate because they do not come to grips with the full meaning of 'potential' in 'potentially a prediction'. I say this because Hempel can always reply that, in cases such as these, even though we do not *have* enough data to predict, *in principle*, we could have predicted. In the evolution case, if the mutation was caused by, say, cosmic radiation, the biologists could very well have predicted the occurrence of a new species. It is simply that this data was not available. But this alone does not entail (i) is false.

> Yet, even to account for the extinction of the dinosaurs, we need a vast array of additional hypotheses about their physics and biological environment and about the species with which they had to compete for survival. But if we have hypotheses of this kind that are specific enough to provide, in combination with the theory of natural selection, at least a *probabilistic* explanation for the extinction of the dinosaurs, then clearly the explanation adduced is also qualified as a basis for a potential *probabilistic* prediction. (my emphasis)[34]

The same thing can be said of the collapsed bridge example.

Even though Hempel's answer appears to save (i) from the above counterexamples, there is something about explanation and prediction that he overlooked in his reply. According to D–N and the resultant structural identity hypothesis, not only must an explanation afford a potential prediction but the *nature* of the explanation must correspond to that of the prediction (= inference based on initial conditions and laws of temporal succession). Thus, if the potential prediction is probabilistic, so is the explanation; if it is deductive, so is the explanation. But we cannot have combinations such as deductive explanations and probabilistic predictions or vice versa, for this would mean that the logic of explanation and prediction would then differ in these cases.

Are there such cases? I say 'yes', and in such a way as to lead to the downfall of (i). All we need is a clear-cut case where the only type of prediction available to the scientist is probabilistic while the explanans is not. The way I will do this is to make use of the fact that not all probabilistic systems afford us with explanations and predictions of the occurrence of a *single* event. Suppose we develop a probability calculus for handling coin-flipping experiments. The probability of getting heads on a single flip is 1/2, two heads in a row being 1/4, 3/4 for getting at least one head in two flips, and so on. But notice that even though we can answer the question, 'What is the probability of the next flip yielding heads up?', we can neither predict what the next flip will bring nor explain why the coin ended up heads (or tails, as it may be) on the basis of this probability calculus alone. (We can, of course, predict the behaviour of large ensembles of coins on the basis of our calculus.) Now in the case of the coins, we have a nice back-up system since we can appeal to deterministic laws of mechanics underlying their behaviour. Even though we are ordinarily in a state of ignorance in these situations, potential predictions are available for the individual coin flips. The crucial point is that not all probabilistic frameworks have the luxury of such a back-up system, and it is this feature that makes things impossible for (i).

Let us first construct an imaginary example to illustrate the above, one that is not really so far-fetched, as we will soon see. Suppose we had a gun that shot bullets through a force field at a screen. What is special about the force field is that it is composed of force vectors that change with time in a *completely* randomised fashion.

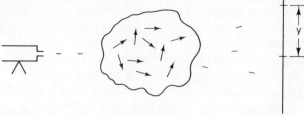

FIGURE 3.3

So we can not predict, in principle, where each individual bullet will arrive at the screen. However, once each bullet makes it to the screen, we have a *post hoc*, causal explanation as to its y-position, one based on the conservation of momentum and energy. Using the analysis of causation presented in the last section, the force field caused the bullet to be deflected to position y on the screen (otherwise the bullet would have hit, say, directly in front of the gun), i.e., a certain amount of momentum was transferred from the field to the bullet during its transit. This is a causal explanation of an individual event but it is *not* a probabilistic explanation at all. I am not saying that the bullet ended up there because it was very likely that it would. On the contrary, it may have been very *unlikely* that it would arrive at that particular location on the screen; nevertheless, it ended up there because the field happened to deflect it there. So, in terms of prediction, we are dealing with a probabilistic inference, but the explanation is not in the least probabilistic but causal.

Notice that I have built into the above case that the behaviour of the randomised field, unlike the coin flips, can not be reduced to underlying, deterministic variables. And it seems that the only way (i) can be saved from the type of example given above is to insist that any randomised system has an underlying deterministic calculus that can yield exact predictions for individual events. But this goes against the tenets of quantum mechanics. For example, we may shoot particles at an atomic lattice, one at a time, in a scattering experiment. If the particle elastically collides with an atom, it will appear at the screen as a scintillation; on the other hand, if it collides inelastically, it will be absorbed by an atom. So, if there is no scintillation at the screen, we explain this by saying that the particle was absorbed. But we could not predict that it would be absorbed on the basis of initial conditions and the laws of quantum mechanics without violating the uncertainty principle (in this case, $\Delta y . \Delta Py \simeq h$, where '$\Delta y$' denotes the uncertainty in our knowledge of position, 'ΔPy', the error in our knowledge of momentum in the y direction and 'h' is Planck's constant). We can not, *in principle*, predict if the particle will be absorbed, for this requires that we simultaneously know its position and momentum within limits of error that violate the above uncertainty relation. The uncertainty relation is a logical outgrowth of quantum mechanics. So, if (i) is to be saved, it would require the downfall of one of the most powerful results of physics as well as the replacement of indeterministic quantum mechanics with a 'deeper', deterministic system. But preserving (i) on the basis of the hope that some day the irreducibly stochastic system of quantum mechanics will be replaced by a deterministic one is a price that any empiricist worth his salt should not be willing to pay.

If the structural identity hypothesis fails, and I believe that the above cases show just that, there are implications for the D–N approach to theories, including the so-called connection between explanation and inference. Contrary to Hempel's contention that explanation and prediction play the same (logical) role in a theory, I believe that a correct analysis of scientific theory should include some sort of a division of labour between the two. I will describe this difference in logic in Chapter 7. In relation to this, I also do not believe in multiplying the *kinds* of explanation there are in terms of the different types of inference. Specifically, contrary to Hempel, I do not believe that there are two different kinds of explanation corresponding to deductive and inductive inference or prediction. I simply hold that because predictive inferences, whether they are deductive or inductive, are separate from explanation, one can not use the former to determine the latter. This, too, will be covered in Chapter 7. The upshot of all this, so far – and it should not be surprising by now – is that there is more to a scientific theory than a body of inferences or potential predictions. If this is so, it is natural to ask wherein lies the 'more'. Various answers to this question are to be covered in the remaining chapters.

8. MODELS

One answer to the above question that has become quite fashionable in some circles in the philosophy of science is the model approach, whose major proponents have included Norman Campbell and Mary Hesse.[35] The model theorists maintain that D–N is incomplete and propose that it be 'modified and supplemented' by its incorporation with models.

The above approach to theories is a contextualistic one, namely, it is not the logical syntax of a theory that is responsible for its explanatory power but the context in which the formalism occurs as supplied by the model that is attached to it. Although it is readily admitted that the formal calculus of a theory is primarily responsible for the theory's predictive power, its explanatory power comes from an entirely different source. Of course, this is a rejection of the structural identity hypothesis as well. A scientist may have a formalism that generates a complete set of predictions but this formalism will nevertheless lack explanatory power if models are absent. This is because, they maintain, the formal aspects of a theory do not contain the *meaning* of the theory's language. The language of a theory, then, is more than just logical syntax plus reference; in a way, scientific discourse is 'too rich' for the formalist tools of analysis in the same way that a formal analysis of poetry would be equally lacking. The language of science is packed with analogies and metaphors, and it is this aspect of scientific discourse that yields scientific insight and understanding – the very aspect that the formalists dismissed as being not notoriously vague! Not only are scientific theories supposed to enable the scientist to organise data and predict; they should also serve as guides for *picturing* nature and how nature works, even if what is pictured can not possibly be observed. Models, not formalisms, are an essential source of these pictures of nature and how nature works. This notion of a model as a picturing device should be contrasted with the more technical notion which occurs in symbolic logic. Here, a model is a set of objects that satisfy the axioms of a formal system; in doing so, these objects become an 'interpretation' of the system. Thus, if geometry is strictly axiomatised, it will be found that 'point', 'straight line', etc., are those objects that satisfy these axioms. In this sense, any interpretation of a formal system is a model. But the scientific use of 'model' intended here goes beyond the logician's meaning.

The logical and scientific uses of 'model' do however have this in common: the relationship between the model and the system modelled is that of similarity. The two systems have a common structure which is technically referred to as a *structural isomorphism* between them. So, for example, a toy airplane serves to model the genuine item in that there exists a structural isomorphism between them with respect to their overall geometrical features. Computers can be used to model or simulate the behaviour of various physical systems because the laws governing their behaviour are isomorphic to the modelled system. As I mentioned, scientific models go beyond this isomorphism. Let us examine how they differ.

Campbell argues that theory not only predicts but supplies an interpretation of the laws used to predict. According to Campbell, a theory contains two kinds of laws: those that express mathematical relations among theoretical variables and those that relate observable variables. These two systems are linked by expressions referred to as *dictionaries*. The dictionary of a theory, then, formally connects the unobservables to observables. Campbell uses the behaviour of gases as an illustration. On the one hand, the laws of statistical mechanics describe the microscopic behaviour of swarms of Newtonian atoms while the Boyle–Charles law relates the pressure, volume and temperature of a gas. One can then derive a relationship between the temperature of a gas, an observable, and the mean kinetic energy of the swarm of Newtonian atoms, which is unobservable. And we can do this with other macroscopic observables such as, for example, 'entropy'. So far, we have two deductive systems whose respective terms are linked by a dictionary. We are still on the formal level of the theory, and we have yet to reach a genuine understanding of gas behaviour in spite of these logical connections. Specifically, we have not reached an understanding of the relationship between pressure, volume and temperature as embodied in the Boyle–Charles law, $PV = nKT$. This is where models come in. The modelling phenomenon is something we all understand and can picture. In this case, we have this image of a vast array of Newtonian atoms whizzing about in a container, bouncing back and forth off the container's walls. One of the properties of this swarm is its mean kinetic energy. Now there is a definite isomorphism between the Boyle–Charles law, $PV = nKT$ and the mechanical law relating the pressure exerted by the swarm on the container's walls, its volume and the mean kinetic energy of the Newtonian atoms, i.e., $PV = c <$K.E.$>$ (c times mean kinetic energy). Thus, the temperature of a gas is very much like the mean kinetic energy of the atoms whizzing about the container; gases are very much like the above collection of Newtonian atoms. This is how we are to picture gases, and this is how a completely unfamiliar phenomenon becomes understood in terms of a familiar one. By means of the above analogy, the language and properties of the model are transferred to the system modelled, enabling us to think of the unfamiliar in terms of the familiar. More important, *new properties* are ascribed to the unfamiliar phenomenon by means of its analogy to the familiar system. This is where the scientific sense of 'model' parts with the logician's use of the term.

Models explain, then, by redescribing the unfamiliar in terms of the familiar. The analogy provides for a redescription of the unfamiliar phenomenon in *new* and '*richer*' terms. Prior to the model, our descriptive repertoire was highly limited. But analogies and metaphors involve the use of language in new and unexplored areas. When we say, for example, that 'Richard fights like a lion', the term 'lion' is applied in an entirely new context. If the analogy is successful, and not all analogies are successful, we have an expanded use of the term. It is in this way that scientific models expand our theoretical vocabulary *vis-à-vis* a phenomenon. Before developing the above analogy between gases and swarms of atoms, we had no basis whatsoever to include 'mean kinetic energy' in our description of a gas. Without the analogy, however, we could neither think of gases in mechanical terms nor could we explain the Boyle–Charles law. With our model, it is readily explained. For instance, why does the pressure of a gas increase when the temperature is increased and the volume held constant? Increasing the temperature increases the mean kinetic energy of the *gas* which, in turn, makes the molecules hit the walls of the container harder, transferring more momentum to the walls which is manifested by an increase of pressure. Even though we can not observe these molecules, we can certainly picture what goes on in a gas when its temperature is increased. This, according to Campbell, is what scientific understanding is all about.

Let us briefly examine some other cases. One interesting thing scientists discovered about gases is that they absorbed and emitted light at characteristic frequencies. These absorption and emission spectra served more or less as 'fingerprints' for identifying the substance in question. Not only that, the scientists were highly successful in developing equations that captured the relations between the various frequencies, wave lengths and spacings among the light spectra of gases. Like the Boyle–Charles law, although these laws of spectroscopy were well confirmed, no one knew what they indicated about gases. Niels Bohr once referred to them in analogy to 'the lovely patterns on the wings of butterflies; their beauty can be admired, but they are not supposed to reveal any fundamental biological laws'. However, Bohr resolved the mystery of the equations of spectroscopy when he presented his famous model of the atom. Prior to this model, the Newtonian atoms of which gases were comprised were naturally thought to be extremely small, dense and irreducible. With Bohr's model, we learn that these ideas are merely an approximation of the properties of atoms, that the structure of the atom is far more complicated. Instead of thinking of the atom in the above fashion, now think of it as being like a miniature solar system. Like the solar system, the atom has a centre or nucleus with orbiting satellites. Each satellite has an energy associated with its orbit; energy is required to transfer it to a higher orbit while energy is released when its orbit is lowered. This is basically how the Bohr model of the atom explains the hitherto mysterious results of spectroscopy. The unfamiliar behaviour of gases is explained by redescribing them in terms of his solar system model: the light given off by the gas is a manifestation of the change of energy levels or orbits of its electrons and because each gas has its characteristic orbital distribution of electrons, it emits a unique set of energy frequencies. In this way, the language of one scientific field, namely mechanics, albeit one modified by a quantum mechanical postulate, has been successfully transferred to another area, yielding a powerful explanation by means of a model.

Another illustration of the explanatory power of models can be found in psychology. One of the more intriguing questions confronting a philosopher or psychologist is 'What is thinking?' The reason why it is such a difficult question is that techniques such as introspection while one thinks or observation of the behaviour of a thinking individual do not seem to lead to fruitful answers. (Ludwig Wittgenstein pointed this out in his lectures on philosophical psychology.) This is why many philosophers and cognitive psychologists turn to computer simulation or artificial intelligence models in order to describe human thought processes. Computers are used to model intelligence first to develop the notion of an abstract, computational system which takes an input 'signal', processes it and converts it into information. Sometimes the input is a computational problem, and, in order to give the answer to the problem, the system must pass through a series of computational or abstract states. So, for example, in order for, say, a digital computer to answer 'How much is 2 + 2?' it may perhaps have to pass through several digital states before '4' reads out on its display panel. Although such computational devices are called *Turing machines*, they are really abstract systems which may or may not be realised by a physical system. For this reason, they are also known as software programs or blackbox systems. Now human beings are to be likened to highly complex Turing machines according to computer simulation models of intelligence, and just as a Turing machine may pass through several computational states between input and output in accordance with a program, human Turing machines have information-processing programs which mediate between stimuli and behavioural responses. The task of the cognitive psychologist is to 'guess' how human Turing machines are programmed, i.e., a human cognitive system is very much like the software design of a computer; thinking is a kind of information-processing; understanding a sentence is like decoding a signal while uttering a sentence resembles the process of encoding a signal; perhaps consciousness resembles the ability to construct an internal model of how one processes information under various constraints; and so on. Of course, the psychologist will opt for that model which best predicts and explains the cognitive behaviour of human beings. For example, cognitive psychologists have attempted to develop programs that explain and predict problem-solving behaviour by postulating that a certain kind of information-processing goes on 'inside' a human subject when placed in a problem-solving situation; and, on the basis of such a model, statistical predictions are made for individuals who are confronted with the same problem. Likewise, programs have been developed to simulate abnormal behaviour such as paranoia in order to gain insight as to how such individuals think or process data in contrast to a psychologically healthy person. Admittedly, all of this is a bit sketchy in comparison to other, established models in science while the use of information-processing models as devices in order to understand intelligence is relatively new; we have yet to see to what extent such models will lead to scientific progress in the behavioural sciences.

The model approach to explanation has been subjected to sharp criticism by the logical positivists and the formalists. In the first place, there is always the danger that the analogy will be taken too literally when all aspects of the model end up being transferred to the system modelled with sometimes ludicrous results. If atoms resembled miniature solar systems in all respects, then electrons would be expected to have craters and satellites of their own as do the planets. So one problem for the model theorist is to supply objective criteria for filtering the positive from the negative aspects of the analogy.[36] The critics claim that such criteria are non-existent. Another point of contention concerns the truth status of theories. Are metaphorical or analogical statements true or false in the same way that literal ones are? Some believe that they are not, that whatever truth they do have, it is poetic in nature and not by virtue of any 'correspondence to the facts'. But, if anything, theories should be true or false in the literal, objective sense, not by being hidden behind some sort of poetic veil. If analogies or metaphors are the essence of scientific explanation, then how can we say our accounts of nature are true?

Even more important, however, there is reason to believe that the model account of explanation may, in fact, be circular. Models supposedly explain by redescribing the unfamiliar in terms of the familiar. But being familiar with something does not necessarily mean that one understands it. So how can we use something we do not understand, although we may be familiar with it, as a model to explain another phenomenon we do not understand? If by 'familiar' we include 'something we understand' then using models to characterise explanation would be guilty of circularity, if not a vicious regress: understanding and explanation, by hypothesis, require a model; and, if the model phenomenon is understood, this means that it has a model of its own, and if that model is understood, it too has a model, and so on. This is why many critics of the model approach to explanation believe that models are not really an essential aspect of scientific theories, that they only play a role in the early, developmental stages of a theory but once the theory is completed, the underlying analogy can be discarded as excess baggage. It may be true that models can generate new explanations and ways of looking at phenomena but once the new and richer language is brought to bear, the model used to get us there can be cast away. So, in the case of the Bohr model of the atom, once we apply the solar system terminology to the atom, we no longer need to compare atoms to the solar system or anything else for that matter. We simply retain this way of looking at things and forget what helped us look at things this way. After all, is it the fact that the properties of the model have been transferred to the modelled system that yields the explanation or is the explanation simply a result of redescribing the explanandum phenomenon in new and richer terms? If the answer is the latter, who cares from whence these new properties come?

The issues of circularity and dispensability actually hinge on the feasibility of a distinction the model proponents make between 'formal' and 'material' analogies. The reason why this distinction is made is because not all models are explanatory. Two systems may resemble each other in their behaviour in that their modes of representation are structurally isomorphic, yet we could not use one to explain the behaviour of the other. Such systems only exhibit a *formal* analogy. For example, the equation of motion that describes a swinging pendulum has the same form as the equation of an oscillating electrical circuit but it is not at all clear that we can use pendulum behaviour to explain electrical circuits although they succeed in mimicking one another. Formal analogies, then, are not what the model proponents have in mind when it comes to explicating the scientific notion of a model. But the distinction between formal and material analogies can not be made by saying that the former are not explanatory while the latter are if we are going to use material analogies to capture the essence of explanation. This clearly would be circular. However, there is a way to make the distinction between formal and material analogies, a way, I suspect, that will lead back to the above dispensability of models difficulty. In the material analogy case, not only is there an isomorphism of behaviour, the key properties of the model are *replicated* in the modelled system. As I pointed out above, the scientific use of 'model' goes beyond that of the logician's by transferring properties from the model to the system modelled. But then explanation comes about by ascribing the new and richer set of properties to the phenomenon; once so ascribed, we can cast away the model used to call them to our attention. So if the above distinction between formal and material analogies is made in such a way as to avoid circularity, it nevertheless allows for the dispensability of models.

This is why some philosophers of science have felt that models are misleading, for it is not the model that captures the essence of scientific explanation but the new and richer meanings – albeit they may be generated by models – that are used to redescribe the explanandum phenomenon. These philosophers feel that explanation is not a matter of inference or redescription in terms of a model but one of semantics, that in some sense explanations come about by redescribing things in 'richer terms'. Of course, the sense of 'richer' here has yet to be explicated. This approach to scientific theory is developed in Hanson's philosophy of science – to which I now turn.

NOTES AND REFERENCES

1. C. G. Hempel, *Aspects of Scientific Explanation* (New York: Free Press, 1965) pp. 189–222.
2. Ibid., pp. 376–403; Hempel, *Philosophy of Natural Science* (Englewood Cliffs, NJ: Prentice-Hall, 1966) pp. 51–69.
3. Hempel, *Aspects of Scientific Explanation*, pp. 364–76.
4. Ibid., pp. 245–51; *Philosophy of Natural Science*, pp. 51–8.

5. Hempel, *Aspects of Scientific Explanation*, pp. 367–76; 406–9.
6. Ibid., pp. 231–44.
7. Ibid., pp. 297–330.
8. Hempel, *Philosophy of Natural Science*, pp. 101–10.
9. Hempel, *Aspects of Scientific Explanation*, pp. 347–54.
10. Hume, *Enquiries*, p. 76. 'Causal relation means predictability . . . It means predictability in the sense that *if* the total previous situation had been known, the event would have been predicted.' R. Carnap, *Philosophical Foundations of Physics*, p. 192.
11. What is of special importance here, something which has been often pointed out, is that Hempel's analysis of theory and explanation depends on an independent, formal characterisation of laws, and I am questioning if this can be achieved in the formal mode. Since Hempel's work, many new, powerful formal systems have been developed in order to capture the logic of counterfactual conditionals. I discuss these systems in Chapter 9. It is my contention that while these systems have led to much progress in solving the problem of distinguishing laws from accidental universal conditionals, they do so in a way that will not help Hempel meet the above criticism.
12. Hempel, *Aspects of Scientific Explanation*, pp. 173–222.
13. What Carnap does to avoid the above paradoxical result is to juggle the form test sentences take on: $C_1 \Rightarrow (T \equiv R_1)$ and $C_2 \Rightarrow (T \equiv R_2)$ and $C_3 \Rightarrow (T \equiv R_3)$ . . . which reads, if we give the subject an IQ test, he will be intelligent if and only if he scores high, etc. These are called bilateral reduction sentences. This way, if he is not placed under test conditions we have no truth functional reason to ascribe intelligence. Unfortunately for Carnap's reformulation, there are still problems. Suppose we felt that, for strong scientific reasons, a particular bilateral reduction sentence is false. Asserting it is false amounts to saying that the experiment *has been* performed with negative results and the system has the theoretical property! (This is because denying material implication is logically equivalent to asserting the truth of the antecedent of the implication and the negation of its consequent.) This is equally paradoxical.
14. Hempel, *Aspects of Scientific Explanation*, p. 205.
15. W. Dray, *Laws and Explanation in History* (London: Oxford University Press, 1957).
16. M. Scriven, 'Explanations, Predictions and Laws' in the *Minnesota Studies in the Philosophy of Science*, vol. III (H. Feigl and G. Maxwell, eds) (Minneapolis: University of Minnesota Press, 1962) pp. 170–230.
17. This case will be discussed in Chapter 7.
18. Hempel, *Aspects of Scientific Explanation*, pp. 363–4.
19. If the reader is interested in a more detailed version of this argument, see my 'Explanations Without Laws', in vol. LXVI, no. 17 of *The Journal of Philosophy* (1969) pp. 541–57.
20. In 'On the Grammar of "Cause" ', *Synthese*, no. 22 (1971) p. 417, I pointed out that transitive verbs like 'cause' and 'makes' are 'dimension-words', i.e., they serve as the most general and comprehensive term for a class of terms of the same ilk. Scientists rarely use 'cause' or 'makes' in their accounts of phenomena but more specified verbs such as 'push', 'pulls', 'dumps', 'oxidises', 'hinders', etc., which all entail that something caused or made

something else happen. The scientist and layman tend to substitute, whenever possible, other words for 'cause', not because that concept is somehow vague or misleading, but because it is more enlightening or descriptive to use a more specific term.

21. Note that while many philosophers disavow the tenets of logical positivism and are quite ready to defend the existence of theoretical entities, they shy away from the view that there are causal connections or powers. Frankly, I consider their refusal to grant an equal ontological status to causal connections to be one last vestige of the positivistic philosophy of science.

22. The arguments I am about to present in favour of characterising the causal relationship in stronger terms than correlation can be found, for the most part, in my 'On the Grammar of "Cause" ' and in my 'The Legacy of Hume's Analysis of Causation', in vol. 2, no. 2 issue of *Studies in the History and Philosophy of Science* (1971) pp. 135–56.

23. Douglas Gasking, 'Causation and Recipes', *Mind*, vol. 64 (1955) p. 479 and p. 480.

24. Feynman, *Lectures on Physics* (Reading: Addison-Wesley, 1963) p. 9–1.

25. What happens in cases where the causal relation appears to be the opposite of what one might normally expect? Suppose a motor scooter runs into a fixed wall and it is stopped, what did the wall transfer to the scooter? The answer is that the wall transferred enough momentum to the scooter to nullify the momentum the scooter had before the crash. Actually, there are *two* causal relations here, although they are not ordinarily recognised: the wall stopped the scooter *and* the scooter caused the wall to get hotter by transferring its kinetic energy to the wall's molecules. Static cases can be handled in like manner. Consider a lead ball resting on a cushion. Two causal relations take place here, which amounts to a momentum exchange between the ball and the cushion, with the net result that neither moves. Or, imagine, if you will, an iron ball being suspended in the air by a magnetic field. The field is actually giving up energy to the ball at a definite rate of transfer.

26. For this reason, a Humean could not conceivably account for what goes on in a momentum exchange between two objects, for a phenomenal language of sense data or impressions is not rich enough to keep track of the identity of quantities and things in these types of physical interactions. See Aronson, 'The Legacy of Hume's Analysis of Causation', pp. 143–5.

27. John Earman, 'Causation: A Matter of Life and Death', *The Journal of Philosophy*, 73 (1976) pp. 23–5.

28. Beauchamp and Rosenberg, *Hume and the Problem of Causation* (New York: Oxford University Press, 1981) pp. 208–11.

29. My answers to the above criticisms and to other, more technical ones which have been introduced in the literature are taken up in my 'Untangling Ontology from Epistemology in Causation', forthcoming in *Erkenntnis*. Two things have been emphasised in that article which should be mentioned here. The transference model is only talking about fundamental or rock bottom interactions. More complex, macroscopic cases of causation are to be analysed as combinations of microscopic transferences. Secondly, I am not at all claiming that '*a* causes *b*' *means a* transfers a quantity to *b*. This is not an analysis of causal locutions but an ontological model which is

intended to make sense of how causation takes place, especially how the causal relationship can be stronger than correlation.

30. In a way, the above picture is misleading because it makes it appear that these forces are transmitted like little billiard balls colliding. According to such a picture, the only effects would be one thing repelling another. But *attractive* forces take place in the same way, namely, by the exchange of a 'force'-carrying particle. What this means is that the old push–pull paradigm of causality is replaced by the above transference model.

31. Max Jammer, *Concepts of Mass* (New York: Harper & Row, 1962) p. 226.

32. See Aronson, 'The Legacy of Hume's Analysis of Causation', pp. 152–6.

33. Hempel, *Aspects of Scientific Explanation*, pp. 367–76.

34. Ibid., p. 370.

35. These treatments can be found mainly in Campbell's *Foundations of Science* (New York: Dover, 1957) pp. 119–58; Hesse's *Models and Analogies in Science* (University of Notre Dame Press, 1966) and *The Structure of Scientific Inference* (Berkeley: University of California Press, 1974) pp. 197–222.

36. See Hesse, *Models and Analogies in Science*, pp. 57–129.

4 Hanson on the nature of a scientific theory

1. PERCEPTION, KANT AND WITTGENSTEIN

Probably the two foremost philosophers Hanson had in mind when he formulated his gestalt approach to theories were Immanuel Kant and Ludwig Wittgenstein. One of Kant's major concerns was the nature of experience, in particular, describing the role the mind played in the having of experience. Contrary to his empiricist predecessors (Locke, Berkeley and Hume), who believed that experience consisted simply in having sense impressions (or sense data) which were organised by the mechanism of association, Kant maintained that this was not enough, that the mind plays a much more active role by synthesising or putting together these disjointed, unrelated elements of experience to form a unified whole.[1] Here we have the beginning of the gestalt approach to perception which Hanson eagerly takes up in the case of scientific observation and where synthesis is an essential ingredient of experience. How does the mind organise its contents? By the application of concepts. This is beautifully summed up in Kant's famous dictum that 'concepts without percepts are empty; percepts without concepts are blind'.

In order to explore further the above notion, consider this analogy between understanding a sentence and perceiving. The words of the sentence correspond to isolated sense impressions while the content or meaning of the sentence to that of experience. In the same way that understanding a sentence cannot be equated with hearing words, experience is not to be identified with having sense data. If, for example, the hearer forgot each preceding word in the sentence string, even though each word in the sentence was picked up by the listener, the meaning of the sentence could not be grasped. On the contrary, each word of the sentence must be 'held in mind', and the sentence string must be processed phonetically, syntactically and semantically in order to draw out its meaning, i.e., to be understood by the listener. The same is said of experience. The 'perceptual units' of experience occur to us over the course of space and time and they too must be held in mind, processed or put together to form a unified whole. This is partly what Kant meant when he wrote, 'All the contents of my consciousness are bound up in a unity.' Again this is done by the application of concepts to our sense data. Suppose we were to have a series of square-like, coloured impressions occur to us in a repeated fashion of red, yellow, blue and orange. How might we synthesise them into a unity by the application of concepts? Well, if we were to think of these sense impressions not simply as the occurrence of isolated, coloured square sense data but instead as *representations* or manifestations of something else, say, some physical process, they can readily be related on this different level of organisation. One obvious way would be to think of them as the appearance of a rotating cube whose sides are serially coloured red, yellow, blue and orange. Such organisation of our sense impressions, however, would require the concept of a moving object in three-dimensional space, and it is Kant's (and Hanson's) contention that our experiences require the application of such concepts on all levels, even if we are completely unconscious of them at work in perception. The role of concepts, in particular theoretical ones such as 'electrons', 'positrons', 'Freudian slips', etc., is especially crucial when it comes to scientific observation. Contrary to the logical positivists, who maintained a view of perception much akin to their empiricist ancestors, scientific observation is not raw data or given but the result of a highly sophisticated, conceptual processing of instrumental readings.

A similar role given to concepts in having certain kinds of experiences can be found in Part II of Wittgenstein's *Philosophical Investigations*.[2] In these passages, Wittgenstein was deeply concerned with this problem of accounting for the fact that certain individuals are not capable of perceiving optical illusions. One example that he takes up is Jastrow's duck–rabbit (Figure 4.1), a case where a figure is seen ambiguously, one very much like Hanson's bird–antelope drawing, which can be found in his *Patterns of Discovery*.[3] Take someone who can not ambiguously perceive such a figure. What went wrong? Wittgenstein refers to the inability to perceive illusions as 'aspect blindness'. It is not at all satisfactory, he maintains, to explain this blindness in physiological terms in the same way we account for, say, colour blindness. There is certainly nothing wrong with the eyes of aspect-blindness victims. Physiological explanations are inappropriate because 'our problem is not a causal but a conceptual one'. What Wittgenstein is alluding to is that the inability to see the above figure *as* a duck or *as* a rabbit reflects certain limitations of the individual's conceptual repertoire. He may, for example, lack the prerequisite concepts that enable him to see the figure in ambiguous ways. (Wittgenstein and Hanson call this type of seeing, *seeing as*.) By 'conceptual repertoire', it is not simply meant that the individual can tell us what 'duck' and 'rabbit' mean, but there is a vast, interwoven system of training, knowledge, skills and even human needs which is associated with the use of the concept in question, and which is responsible, in part, for shaping the individual's experiences. (Wittgenstein refers to factors like these underlying the application of our concepts as *forms of life*. This notion will be discussed in the next chapter.) So, an account of the difference between how things are perceived in *seeing as* cases involves the use of concepts, and, once more, we find that there exists an inextricable link between language and experience.

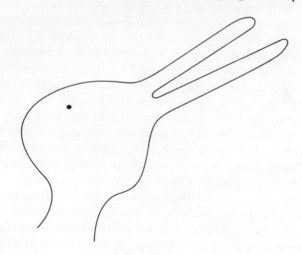

FIGURE 4.1

2. THEORY AND OBSERVATION

Hanson uses the above interdependence between concept or language systems and experience as a basis for his attack against the positivistic and Hempelian approaches to scientific theory. In the first place, Hanson maintains that perceptual ambiguity also occurs in the sciences; and, in a way not very unlike the duck–rabbit case. As we have noted, the logical positivists held the inductivist approach to theories; theories were simply logical constructions of the facts. But what are the facts? Are they really 'hard, plain and unvarnished'[4] as the positivist would have it? Are they incorrigibly and indisputably perceived by the scientist? History seems to tell us otherwise in light of the multitude of disputes over various interpretations of experimental results. On the contrary, 'reading off' the experimental facts is indeed a very delicate conceptual matter, a task which does not occur in isolation but, instead, requires the scientist to exercise precise judgement. In the same way that the correct interpretation of an element in a drawing requires that it be related to the other elements by means of a system of concepts, the interpretation of experimental data likewise involves the application of theory. There are no such things, then, as isolated facts. A fact is such only to the extent it fits within a pattern or a total picture of things.

Even though Hanson does not identify seeing with *seeing as*, he feels that the latter can serve to 'illuminate' the former by emphasising how observation and theory are so tightly inter-woven.[5] But if our theories in some sense determine what we ob-serve in that observation involves the application of a scientific conceptual framework to seemingly unrelated data, then it is possible for two scientists who share the same set of data to arrive at opposite observational interpretations because they maintain (and apply) opposite theories in the first place. (This exactly parallels the duck–rabbit case where two individuals see the same figure but one sees it as a duck while the other sees it as a rabbit.) Hanson is quite aware of and welcomes such a possibility. Things are not as cut and dried with scientific observation as certain 'tough-minded' philosophers would have it. One aspect of science to which the logical positivists and formalists were completely blind is perceptual relativity: the same set of data can lend itself to a variety of interpretations (or facts), depending on what theory is applied.

Hanson introduces the above doctrine by having us imagine Johannes Kepler and Tycho Brahe together observing the dawn.[6] It is true that in one sense of the word 'see', the two astronomers see the same thing in that they are looking at the same objects: an orange ball, the Earth's horizon and their relative motion. But when it comes to seeing in the sense of scientific observation, they do not see the same thing; their visual experiences are quite different because each scientist brings to bear his own astronomical system to the dawn: Tycho Brahe is observing the sun moving above the horizon as it makes its way around the Earth, the centre of the solar system, while Kepler observes the Earth turning into the light of a stationary sun. As it was in the duck–rabbit case, the elements of their respective visual experiences may be exactly the same but they are patterned or gestalted differently because different conceptual schemes are doing the organising. So perceptual relativity is a natural outgrowth of Hanson's theory of observation. The facts, then, are not 'out there' in some absolute sense but are relativised to the conceptual scheme of the observer. Not only that, the individual who has more theoretical concepts at his disposal can actually see or observe (= pattern data) more things than a less fortunate individual. Two individuals may be looking at a slide under a microscope. While one may see just a group of cells the other sees a rapidly growing cancer; although each person saw the same group of cells, the latter individual 'saw more', and this is so because said person possessed more medical knowledge.

3. THE GESTALT APPROACH TO THEORY AND EXPLANATION

Throughout Hanson's writings we find examples of drawings that are seen one way and then another, faces that pop out of blotchy figures, shapes that change their meaning as we switch from one visual context to another, and so on. What is the point of all these exercises in illusion, especially since they appear to be more relevant to aesthetics than to the philosophy of science? The answer is that they are to enlighten us about the nature of scientific understanding if not understanding in general. They also lead to a highly controversial claim about the relationship between understanding and observation: understanding is actually a species of perception.[7] In other words, not only are Hanson's gestalt examples intended to instruct us on the nature of scientific observation, the very same approach applies to scientific understanding.

Among the things his gestalt examples are meant to show about scientific understanding are these. In the same way that the background of each figure serves as a context which accentuates certain of its elements and makes their meaning explicit, theories serve as backgrounds which enable the scientist to pick out of a vast array of data those facts that possess explanatory relevance. (Imagine a speck of dirt on a drawing of a face. We would take the 'dot' to be incidental to the other lines because of our interpretation of the total figure.) Suppose there are several sets of initial conditions and laws that lead to the deduction of an explanandum. How are we to select our explanans from these sets? Hempel's answer is that the principle of simplicity determines our choice of explanans.[8] But Hanson would quickly reply that even our notion of the simplicity can not be abstracted from theory since what may be simple according to one theory may be highly complex in another. This is why explicating simplicity in terms of the 'smoothest fitting curve through points' is a charade: the smoothest curve in one system may not be so in another. Theory, then, must glean out which data are relevant to explain a given phenomenon. In relation to this, unlike the D–N view of explanation which allows for explanations of events to take place in isolation from one another – i.e., so long as each explanans has its own set of laws and initial conditions – each explanation, according to Hanson's story, actually has an entire conceptual framework packed into it in the same way that the parts of a picture get their meaning in terms of the whole. Finally, in the same way that we can speak of these gestalt examples having various levels of organisation, the language of science is semantically multi-levelled. This is the key to Hanson's notion of theory-loadedness and his analysis of scientific understanding.

Hanson's analysis of theory and explanation has, among other things, logical syntax replaced with semantics. Recall that, with D–N, laws of nature were the basic explanatory connecting links among events; they are usually statements of the logical form 'whenever A then B'. The job of explaining is not carried out by the logical syntax of the law, however, but through its conceptual or semantical power which is carried by its theoretical terms. Exactly what does Hanson mean by this? In the first place, the language of science is many-levelled, and the basis for determining each level lies in what I will call *conceptual connectivity*.[9] We find that there are more connections among the concepts on some language levels than others, ranging from a sense datum level where there are no connections among sense datum terms to the most highly sophisticated theoretical levels where there exists a myriad of conceptual links. Notice right away that even though statements of the universal conditional form logically connect the antecedent and consequent terms by means of a universal quantifier and material implication, Hanson does not count this as a semantical connection; the basis of a true conceptual link lies in the meaning of the concepts and it would be possible to have universal conditionals connecting concepts without there being a corresponding conceptual relationship between them within a single theoretical framework.

> The terms of physics thus resemble 'pawn,' 'rook,' 'trump,' and 'offside' – words which are meaningless except against a background of the games of chess, bridge, and football.[10]

I must reiterate that Hanson is maintaining that the connections among these concepts are above and beyond any logical connections that may or may not exist between them in the form of universal conditionals. This is his most fundamental assumption. The extent of conceptual connectivity will determine the level of each system of concepts. So when Hanson speaks of the semantics in contrast to the logical syntax of language, he means conceptual connectivity and levels of language.

Hanson's semantical apparatus leads to a model of explanation which rivals D–N. Given that we can now speak of a spectrum of semantical levels of language, explanations turn out to be theory-loaded redescriptions of events: one and the same event, e, which is originally described by D, when an explanation is requested, is redescribed by R when the explanation is presented. The relationship between e, D and R is not that of being subsumed under a law (e can be subsumed under a law on the D level alone). While D and R describe the same event, e, R occurs on a higher semantical, more theory-laden level than D. In other words, explanation takes place not by deduction of e from laws and initial conditions but by going from one semantical level to a higher theory-laden level of describing events:

Explanans located on level, $l_n + 1$
Original description on level, l_n
Event to be explained

As an example of l_n and $l_n + 1$ levels:

> Whenever the youngster says 'lightning and thunder,' he probably means 'flash and rumble'. Again, a lot may follow, but what follows for him is different from what follows for the meteorologist – for whom 'lightning and thunder' probably means 'electrical discharge and aerial disturbance'.[11]

Terms on the $l_n + 1$ level are 'richer' and more 'charged'; 'they carry a conceptual pattern with them'. This is why the source of explanatory power in science lies in the semantical structure of theories instead of their logic.

One apparent consequence of the above view of explanation is that 'the more "phenomenal" a word, the less "theoretical" it is',[12] lessening its explanatory force. The least theory-laden level of language takes places on a sense datum level, which is entirely lacking in explanatory power, even if, as I have said, each and every term is connected by a universal conditional. Generalisations of sense data simply do not yield explanations but *only repeat the facts*. Recall the rotating cube example in the first section. We could easily generalise over the repeated pattern of red, yellow, blue and orange square impressions, formulating a statement like this of a universal conditional form: whenever a red-square impression occurs, a yellow one immediately follows. But can we really explain the occurrence, say, of a yellow-square impression by citing the above law along with the initial condition that a red square just occurred? Intuitively, the answer is 'no', that an explanation must resort to a higher-levelled, object language such as 'the red-square and yellow-square occurrences were the appearance of a rotating cube'. Let us look at some more examples of explanation as theory-laden redescription. Suppose a shiny copper penny turns colour to a dull green after being exposed to the air after a period of time. One might ask 'Why did the penny turn green?' Assuming the inquirer has knowledge of chemistry, a perfectly good answer might be 'It turned green because it has been oxidised.' The change in colour is explained by redescribing it on a higher level, in terms of a change in the chemical composition of the penny's surface. Take yet another case, this time explaining human action. You are watching a chess game and one of the players moves a piece in such a way as to expose it to capture. Since this would lead to his opponent's immediate advantage, you are perplexed and ask 'Why that move?' Your friend, who happens to be a chess expert, replies 'He just baited a trap', and sure enough, when the opponent captures the piece, he is checkmated in three moves. The explanandum move is related to other moves in the combination by redescribing it in terms of the language of chess strategy. Consider this last example by going back to the colliding spheres described in the last chapter. This time, the movement of sphere *b* at the time *b* is redescribed as the resultant stage of a process of momentum transfer from sphere *a* at time 2, i.e., to 'motion' we add the higher-level dimensions of 'momentum' and 'energy'; in doing so, we move from a kinematical to a dynamical level of redescription.

I come now to understanding being a species of perception. This result should not be surprising by now because perceiving and understanding are both theory-laden activities. In the case of scientific observation, an isolated act of perceiving involves the entire conceptual apparatus of the scientist in order to gestalt or pattern the data. But the same use of higher-level concepts goes on when it comes to understanding a phenomenon. Thus, Hanson has a structural identity hypothesis of his own: there is only a pragmatic difference between observation and understanding.[13] Hanson appeals to a case history of the discovery of the positron in order to bring this out. His argument is that while physicists had enough data, in the form of tracks in a Wilson cloud chamber, to indicate clearly the presence of a positron (a positively charged electron), they were in no position whatsoever to make such an observation because their theories were lacking. Dirac's theory of the electron had yet to be formulated and its absence left the experimental physicist conceptually unprepared for the particle's discovery. With the advent of Dirac's revolutionary work, they were subsequently able to interpret certain lines in a photograph as the cloud chamber tracks left by a positron. But the very same theory enabled them to understand these tracks as those made by a positron. Here perception and understanding go hand in hand; in each case, 'an intelligible, systematic, conceptual pattern for the observed data' has been supplied.

If Hanson's story is correct, Hempel's structural identity hypothesis must be rejected, for two systems of inference may lead to the same set of predictions but differ in terms of explanation and observation because each has a different system of concepts attached to it.[14] For example, Newton's version of celestial mechanics may have yielded the same set of predictions as Lagrange's treatment of the same subject matter but explanations and observations in terms of gravitational forces do not resemble the least action principles of Lagrangean physics.

Another of Hanson's major departures from the logical positivist and the D–N approach to theory concerns the logic of discovery. Hempel is very careful to distinguish the process of discovering or arriving at a hypothesis from the workings of a theory as a finished product. Discovery is not a process subject to logical analysis in the same way that a theory's explanatory and predictive power are. After all, a scientist may discover a new theory simply by guessing or 'dreaming it up', activities that are of more concern to a historian or psychologist than a logician. Hanson feels otherwise. He claims boldly that the so-called subjective process of discovery actually has a logic of its own albeit it is not that of D–N. In the first place, the inference is made from the data to the hypothesis, which is the reverse of D–N where data are deduced from laws and initial conditions. Along with Charles Sanders Peirce, Hanson refers to this new logical process as *retroduction*:

> Yet every plan of its advance [scientific theory] is first laid by Retroduction alone, that is to say, by the spontaneous conjectures of instinctive reason; and neither Deduction nor Induction contributes a single new concept to the structure.[15]

Although retroductive inferences are neither inductive nor deductive, it does not follow that they are arbitrary or 'a-reasonable' – they are just a different form of reasoning. Hanson goes on to show retroduction at work by examining the reasoning process Kepler went through in the course of discovering that the orbit of Mars was neither circular nor ovoid but elliptical. If Hanson's interpretation is correct, Kepler's reasoning was neither inductive, deductive, nor a combination of the two, but something else.

Before we evaluate the above case study, we must first be extremely cautious about the exact nature of Hanson's claim, especially about what it means to speak of a process of inference being something other than deduction or induction. Roughly, such a claim entails, at the least, that these inferences are in accordance with rules other than the axioms of inductive or deductive systems of logic. Has Hanson shown the existence of these rules in his case study? I think not. Notice that it will not do to reason that because retroductive inferences go from data to hypotheses rather than hypotheses to data (as in the case with D–N) the *nature* of inference in each case differs. Inferences are not characterised by their conclusions but by the rules that govern them. The problem of theory construction might be set up like this: given a set of premisses $P_1, P_2, P_3 \ldots P_n$ (which describe data) and a conclusion Q (which describes an observation we wish to explain), what premiss or premisses (i.e., our hypothesis H) must be added to $P_1, P_2, P_3 \ldots P_n$ in order validly to infer Q? If Hanson is correct, one's reasoning in the course of arriving at H from P_1, $P_2, P_3 \ldots P_n$ and Q will be neither deductive nor inductive (nor some combination of the two).

But one might arrive at H without reasoning at all! One might make a lucky guess and discover that when H is combined with P_1, $P_2, P_3 \ldots P_n$, Q is a valid consequence. On the other hand, if one arrives at H by reasoning and then goes back over the reasoning, one might discover it to be quite logical and even deductive, as illustrated in Hanson's case study of Kepler's discovery of elliptical planetary motion. To quote Kepler:

> My reasoning was the same as that of Chapters XLIX, L, and LVI. The circle of Chapter XLIII errs by excess. The ellipse of Chapter XLV errs by deficiency. And the excess and the deficiency are equal. There is no other middle term between a circle and an ellipse, save another ellipse. Therefore the path of the planet is an ellipse.[16]

But this looks like good old deductive reasoning to me!

The reason Hanson's claim about retroduction being a new form of inference seems to be so unwarrantable is that he never supplies the reader with a model of how discovery takes place, one whose underlying rules are neither deductive nor inductive. Instead, he falls back again on his gestalt approach to theory, which, by now, has become a highly overworked concept, to say the least.

> Peirce regards an abductive inference (such as "The observed positions of Mars fall between a circle and an oval, so the orbit must be an ellipse') and a perceptual judgment (such as 'It is laevorotatory') as being opposite sides of the same epistemological coin. *Seeing that* is relevant here.[17]

Not only are observation and understanding wedded under the patterning or gestalt notion, they are joined by reasoning processes underlying theory construction. In his attempt to unify the activities of perceiving, understanding and reasoning to a hypothesis under a single concept, Hanson has pushed the gestalt approach beyond all reason, even to the point of self-contradiction, for whatever perception (or scientific observation) may be, it certainly is not an inference; if anything, perceiving and inferring (or reasoning) are contrasted to one another, something Hanson clearly accepts in his chapter on observation.

In answering the charge that discovery should be studied by historians and psychologists instead of logicians, Hanson replies that 'the exhaustive and exclusive dichotomy "Psychology or Logic?" may win debates occasionally, but it cannot win the guerdon of truth'.[18] What Hanson wants to do here is to bust the traditional dichotomy between reasoning and psychological processes, arguing that even if the psychological processes at work in the course of discovery may not be in accordance with the rules of deductive or inductive logic, it can be said that they are nevertheless in accordance with *other* rules. There is a right way and a wrong way to go about discovery. To refuse to call this logic because the rules are not the formal rules of symbolic logic is to be guilty of nit-picking. But this makes things even worse for his stance on the logic of discovery, for cognitive psychologists have devoted much research to developing models for problem-solving or theory construction, and what they come up with as rules or guides underlying inferences from data to hypotheses is a highly complex combination of judgement, inductive guesswork and deductive reasoning. Retroduction or abduction, as well as gestalt psychology, play no role whatsoever in these models. Again, we are never told what these rules are. Until we are presented with a rival model of theory construction or problem-solving where novel forms of reasoning are utilised by the problem-solver, Hanson's enterprise remains highly suspect. The interesting thing about this recent research in problem-solving is that a good case can actually be made for a logic of discovery without having to go all the way by postulating new types of logic. All that has to be done is to show that there exists a method or a way of going about theory construction where we can speak of good or bad ways of doing it. Surely, trying to find guides or rules underlying the activity of discovery is a task worthy for the philosopher of science. It is simply that Hanson has gone about it the wrong way.

4. PROBLEMS CONCERNING THE GESTALT APPROACH TO THEORIES

The above difficulties that plague Hanson's attempt to establish retroduction as a new logic of discovery are actually symptomatic of his other claims about scientific theory and observation. What we have here is a case of unfulfilled promise, for his gestalt programme is just that: a programme whose details have yet to be carried out in order for us to be in a position to evaluate the extent of its utility for comprehending how scientific theories work. We are confronted with a claim that there exists a new type of inference but we are never told what it is. His model of explanation is based upon a semantics of language which is never fully brought out. Finally, although the doctrine of perceptual relativity is stated right away, we are really never supplied with adequate criteria that tell us when two scientists are perceiving different things other than they hold different theories. All we are left with are unpacked gestalt analogies with perception. Let me go into the details of my complaints.

As the reader can tell, I find it highly questionable to use the gestalt model as a vehicle for establishing Hanson's claims. Consider his version of scientific explanation being theory-laden redescription. We know by now that theory-ladenness is based upon the notions of conceptual connectivity and levels of language. Until we are supplied with a clear elucidation of these semantical properties, Hanson is open to the charge of vagueness by the formalists, a criticism I believe he deserves. What exactly does it mean for one level of language to be more theory-loaded than another? One thing it can *not* mean: one level is more theory-loaded than another in that the use of its terms entails the truth of a law, an entailment which does not take place on the lower level, i.e., the more theory laden, and more lawlike connections among terms. As J. C. Smart correctly points out, if *this* is how 'theory-loadedness' is characterised, then to say that explanation is theory-loaded redescription offers no real alternative to D–N analysis.[19] It amounts to saying that we are explaining by redescribing events in terms of laws, i.e., the new semantical level now subsumes events under laws which the old level of description did not. But this is D–N all over again. If 'theory-loadedness' is going to work it must be characterised independently of laws.

In spite of Hanson's claim to the contrary, I seriously doubt that conceptual connectivity and, hence, 'theory-loadedness' can be elucidated without bringing in laws. Let me put it this way. If, as he claims, scientific concepts are not connected by laws, then how are they connected or 'patterned'? The best Hanson has to offer in answer to this is an analogy between theoretical terms and 'game-jargons':

> Cause-words resemble game-jargon, as we noted earlier. 'Revoke,' 'trump,' 'finesse,' belong to the parlance system of bridge. The entire conceptual pattern of the game is implicit in each term: you cannot grasp one of these ideas properly while remaining in the dark about the rest. So too, 'bishop,' 'rook,' 'checkmate,' 'gambit,' interlock with each other and with all other expressions involved in playing, scoring and writing about chess.[20]

Recall the problem of filtering out the positive from the negative aspects of an analogy. When it comes to game-language, the rules and strategies of the game determine the connectivity of its concepts. As we have seen, a particular type of move in chess may be related to others in terms of a strategy or manoeuvre such as setting a trap or pitfall for one's opponent. But this kind of conceptual connectivity involves conventions and intentional behaviour; and, surely, we do not want to say that conventions, linguistic or otherwise, govern connectivity among scientific concepts. If so, what would be analogous in the language of science to the rules and strategies found in game-jargon?

The only answer to the above question, I feel, is that laws which express invariable relations between properties are the major vehicle of conceptual connectivity in science. I can not think of anything else that does the job. What else can be meant by 'The concept "force" is related to that of "acceleration" ' except that the properties these terms denote are related in some specific (and universal) way, say, by Newton's second law of motion? (We need not bother with relating concepts by definitions here, for this is not what Hanson has in mind anyway.) We can use this idea to supply a criterion of conceptual connectivity, one which I will make use of later on:

> Two concepts are conceptually related [other than by definition] within a science if and only if there exists at least one law or theorem of that science in which they both appear.

Ultimately, I want to say that conceptual connectivity has a lot to do with the ontology of a science. After all, the reason why 'force' and 'acceleration' are concepts related in Newtonian mechanics is that, according to this system, forces *cause* accelerations in a lawlike way. If *this* isn't the source of conceptual connectivity in science, then the onus is on Hanson's followers to supply another. I frankly doubt that they can.

Not that I think 'conceptual connectivity', 'theory-loadedness' and 'semantical levels of language' are worthless ideas which should be scattered to the winds. I just feel that Hanson has gone about presenting them in the wrong way. Whether the contextualist would be willing to admit it or not, they happen to be *formal* notions which require the very backing of logical and mathematical frameworks we are advised by these philosophers to ignore. The gestalt approach fails to get the above concepts off the ground in that it is too arbitrary and intuitive as a criterion for discerning levels of theory-loadedness. It simply can not do a job which more formalistic approaches can, as I hope to demonstrate later on.

In relation to this, there is yet another question concerning his account of explanation as theory-loaded redescription. We realise by now that this way of looking at things locates the explanatory power of a theory in a hierarchy of language levels as opposed to the logical syntax or lawlike connections on each level. But if scientific understanding does not result from the subsumption of events under laws of nature, how is it that redescribing events on a more theory-laden level explains?

What is it to supply a theory? It is at least this: to offer an intelligible, systematic, conceptual pattern for the observed data. The value of such a pattern lies in its capacity to unite phenomena which, without theory, are either surprising, anomalous, or left wholly unnoticed.[21]

Our only recourse, it seems, is to appeal once more to analogies; this time, between gestalt patterns and theories. This naturally leads us to ask what it is about these semantical levels that somehow patterns events or makes them coherent? And, again, we find the gestalt analogy wanting because it really does not supply an answer to this question; and, just as it was before with Hanson's other major ideas on theory, the notion of making our observations *cohere* or *fit together* by the application of theory, requires formal backing which is conspicuously absent in his treatment. Without this, how can we ascertain *what* data a theory organises and *how* this patterning takes place?

5. PERCEPTUAL RELATIVITY REVISITED

'Do Kepler and Tycho see the same thing in the east at dawn?' Hanson is fully aware of the complexities involved in answering this question. Philosophers have argued about this for centuries and they are debating it now. Hanson's answer appears to be 'That depends on what is meant by "seeing the same thing" ', and he goes on to lay out some distinctions or senses of 'seeing the same thing'. There is a primary sense of 'seeing the same thing' where by this we mean different observers discern the same set of objects. When observers perceive one and the same set of objects from different perspectives, the objects of perception are said to be *public*. Now Hanson wants to say that Kepler and Tycho see the same thing in this sense. Each has public access to the same sun, Earth and their relative motion. I will call this public sense of seeing when different observers see the same thing, see_1.

See_1 is not to be confused with the sense of 'seeing the same thing' when it comes to theory-laden, scientific observation. This second sense is based upon a strict analogy with gestalt or seeing as cases, which Hanson calls *seeing that*. As it was in the gestalt figures which lent themselves to ambiguous interpretations, one and the same set of scientific data are equally amorphous; and, in the same way that the application of concepts resolves the ambiguity in seeing as cases, the scientist's theories determine how the data are to be interpreted, yielding a total scientific visual experience. In that two scientists who see₁ the same things pattern their visual fields differently because they hold different theories, they see that or scientifically observe different things. Hence *seeing that* is relativised to the scientist's conceptual framework. It is this sense of 'see' where theory determines what the scientist observes and where we can speak of one observer seeing more than another.

As I have noted all along, Hanson invokes a simple formula for scientific observation being seeing₁ plus the application of a theoretical system to one's visual field. He is very careful to insist that this latter sense of *seeing that* is a full-blown sense of 'see'. If not, he is open to the criticism that he is not dealing with scientific perception or any kind of perception for that matter, for the logical positivist could easily reply that Kepler and Tycho see the same thing, all right, but *infer* opposite opinions about what it is they are, in fact, seeing because their inferences are based on different opinions. Thus, Hanson's sense of 'see' is no more a case of seeing something than is 'see' in 'I see what you mean', or 'I see what these events imply'. This is why the above formula is so important, for Hanson insists that, when it comes to seeing that, no inference takes place, as it was in the cases of seeing as; the concept of 'duck' was applied simply in order to gestalt the data.[22] In other words, we come right back to the Kant–Wittgenstein thesis that perception and the application of concepts are inextricably tied. The application of a concept in perceptual judgement is not to be confused with an inference from visual data on the basis of laws.

The above thesis linking concepts and perception is highly debatable. There are problems for those who insist that concepts are an essential ingredient of perception. Do children and animals perceive things? For example, 'visual cliff' experiments were performed on newborn animals and babies. They were placed on a perfectly flat surface which abruptly dropped off, forming a 'cliff' in which the animal or baby could fall if it were not for an invisible glass shield which extended out from the plane. When the babies and newborn animals approached the artificially created precipice, they stopped and would not go on. How can we account for their behaviour? The most natural and obvious way to do so is that each saw the cliff and stopped. But if the babies and animals each saw the precipice, what concepts were they employing? Can it be said that they even have concepts in the same way that adult human beings have the topological concepts of 'plane' and 'precipice'? If the answer to these questions is 'yes', then the above doctrine appears to be committed to some form of innate ideas in animals and babies. If their answer is that the animals did not see the cliff but stopped short of it because they were simply responding to visual stimuli, they are open to the charge of begging the question; for if these animals behave the way they do as a result of gaining information about their environment on the basis of visual (light) signals, why not call this seeing? This dispute over the role of concepts in perception can not possibly be resolved here but I do want to call the reader's attention to these two rival approaches to perception: one tradition maintains that perception involves the application of concepts while the other views perception as gaining information by processing or analysing light signals. The latter maintains that human beings and animals can visually discriminate objects without recourse to concepts, theories or beliefs.

It just may be that Hanson has failed to make a very important distinction between *seeing* and *appreciating what it is that one sees*.[23] While the former can take place without bringing in our theories and beliefs there is nothing at all wrong with saying the latter depends on them. For example, a four-year-old child may look at an X-ray tube and see it perfectly without realising that it is an X-ray tube which he, in fact, sees. This would be so, not surprisingly, because of his ignorance of X-ray tubes, as if four-year-olds should know about such things. In other words, not knowing about the object of perception does not imply that it can not be perceived. Suppose a scientist and a layman look at a bluish glowing line in a cathode-ray tube. The scientist knows he is looking at a stream of electrons while the layman does not. Does this mean that this layman did not see the stream of electrons because of his ignorance of electricity and magnetism? Hardly at all. So why not describe the difference between Kepler and Tycho simply in terms of each appreciating what he sees rather than in seeing *per se*?

There are good, logical reasons for making this move. 'Seeing' and 'seeing that' are what Ryle calls *achievement* verbs: seeing *X* or *seeing that X* entails that *X* is the case. In the same way we can not know what is false, we can not see what is not the case. If, while intoxicated, someone claims to see pink elephants when there are none present, even if our inebriated friend is having a visual experience, we dismiss his claim because there are no pink elephants out there to be seen. He simply and mistakenly *thinks* he sees them; he is hallucinating them. Likewise, a scientist can not observe that which is not the case. If so, *seeing that* can not be like *seeing as* in a very important way. It can not be true that Kepler sees the Earth turning into the light of the sun *and* Tycho sees the sun moving relative to a fixed Earth – even if our observers apply opposite theories to the same data. In that *seeing X* entails *X*, they can not both be right. In the gestalt cases, however, we can readily see a figure one way and then another without contradiction because we are dealing with ambiguous appearances and not reality. To exaggerate my point, suppose two observers watch a ship travel on a westward direction on the ocean; one believes the Earth to be flat while the other holds that it is round. Can Hanson seriously claim that the former *sees that* the ship is heading towards the edge of the Earth, *sees that* it is about to fall off the edge?

My complaint about Hanson's analysis of *seeing that* and the subsequent formulation of the doctrine of perceptual relativity is that it is not at all made clear how going from see$_1$ to seeing that on the basis of applying a theory indicates a *perceptual achievement* has actually occurred. The same facts can be accounted for by appealing instead to an inference on the basis of what one sees$_1$. Surely, the successful application of a theory to data does not necessarily mean that a perceptual achievement has taken place. One may see$_1$ *X* and may know that *X* and *Y* are correlated. On this basis one knows that *Y* is present. Does this mean that one *sees that Y*? What if *X* and *Y* are separated by a billion miles or by some opaque barrier? It would be far better to say, in cases such as these, that *Y*'s existence was *inferred* by observing *X* and applying the hypothesis that *X*s and *Y*s are correlated.

What I wish to point out by the above is that the nature of the perceptual achievement in scientific observation differs significantly from the perceptual skill required in the gestalt cases, and that the thesis of perceptual relativity can not be presented in terms of the gestalt model because the analogy between *seeing that* and *seeing as* breaks down. This becomes dramatically apparent in the Tycho–Kepler case in light of the fact that Galileo proved that the movement or lack of movement of the Earth (as an inertial frame) is precisely the kind of thing that Kepler or Tycho could not possibly observe as they look east at dawn. In his *Dialogue Concerning the Two Chief World Systems*,[24] Galileo has us imagine a series of kinematic experiments performed on a ship at rest and then the same experiments performed when the ship is moving at a constant velocity on a very calm sea. Experimentally, there would be no discernible difference between the ship at rest and in constant motion, no matter how fast the rate of the moving ship; if the experimenter were locked inside a windowless cabin he could not tell if the ship were moving or not. This formulation of the *principle of relativity* was one of Galileo's greatest achievements and he used it as a powerful weapon against the opponents of the heliocentric view. They maintained that if, contrary to Tycho's theory, the Earth moves, we should be able to observe that it moves. (For example, if two identical cannons were placed back to back along the supposed motion of the Earth and fired projectiles, the distance traversed by the projectiles should differ if the Earth moved and should be equal if the Earth were at rest. This experiment was actually performed, 'confirming' the geocentric theory. Galileo simply refuted it by showing that when the heliocentric theory is conjoined with the principle of relativity, projectiles should land equidistant whether the Earth was at rest or moving at any velocity whatsoever.) In other words, there are certain things that can not be perceived under specific visual conditions and the absolute motion of the Earth from a vantage point on its surface happens to be one of them.

Suppose we took seriously the claim that Tycho saw that the sun was rising above a fixed Earth and Kepler saw that the Earth was rotating into the sun's rays. Putting the contradictory nature of this claim aside, we ought to ask these questions. *What would it look like for the Earth to turn or the sun to rise above a fixed Earth? How could we put one in position, here, to perceive the difference?* These are the crucial considerations about differences in perception that Hanson neglects. (Substitute 'feels that *X*' for 'sees that *X*' and this point becomes even more obvious.) Surely, it can not be sufficient to put one in position to see that the Earth is turning towards the sun – as opposed to seeing that the sun is rising above a fixed Earth – by bleating out the heliocentric theory and pointing the observer at the dawn. This is where Hanson's theory of observation is entirely lacking. If Tycho and Kepler are to see things differently in any sense of the term, then Hanson must draw out a difference in the *visual aspects of the situation*: it must be shown that *things look, appear* or *behave differently to each observer*; and, as I have argued above, this can not be done in the Tycho–Kepler case because of the principle of relativity – a fixed Earth with the sun moving about it and a rotating Earth into a fixed sun appear exactly the same to someone observing the dawn. Any formulation of perceptual relativity in the sciences, then, must go beyond a mere difference in the theories or beliefs of observers to include how things look, appear and behave.

6. DRETSKE'S ANALYSIS OF *SEEING THAT* AND HIS REFORMULATION OF PERCEPTUAL RELATIVITY IN THE SCIENCES

What any analysis of scientific observation must do is to incorporate theory with the appearance aspects of things. This is the heart of Dretske's analysis of *seeing that*. I am presenting it because I think that it adequately comes to grips with the above-mentioned problems that confront the gestalt approach to observation and lends a clarity to the complexities concerning the relationship between theory and observation which has hitherto been lacking. Dretske's answer is based on the above distinction between seeing and appreciating what it is, in fact, that one sees. Seeing (or see$_1$) is not relative at all while appreciating or identifying what it is that one sees is relative to this particular aspect of scientific theory: any theory must contain statements about how things *look, appear, behave*, etc., in relation to what they are. This is the key to understanding the nature of scientific observation or *seeing that* and to a more reasonable formulation of perceptual relativity.

Dretske calls see$_1$ *non-epistemic seeing*, parting from those who maintain that perception always involves the application of concepts. A sufficient condition for an observer, S, to see$_1$ D, is that 'D is visually differentiated from its immediate environment by S'. By 'visually' here is meant 'that S's differentiation of D is by visual means, in terms of D's looking some way to S instead of D's feeling or tasting a certain way to S'.[25] For example, S sees$_1$ a tennis ball, involves S visually differentiating something that looks round, fuzzy and appears to be darting about with respect to, say, a green background of S's visual field. But S's visual achievement does not require that he *knows* that a tennis ball has been picked out. If he didn't know it, does this mean he didn't see it (or even hit it with a racket)? 'Total ignorance is not sufficient for total blindness.'[26] Seeing$_1$ takes place, then, whether or not S 'exploits his visual experience in any way whatsoever', even whether or not S applies any concepts or theories to D. Like the child who sees the X-ray tube, S may not have *any* concepts or theories at all about D. Likewise, in order for a child to see a dromedary at the zoo must he possess the corresponding concept or some kind of zoological theory? On the contrary, see$_1$ is logically independent of our theories and beliefs.

Appreciating what it is we perceive, however, does entail things about our theories and beliefs. This is where *seeing that* comes in. Realising, identifying or being informed about the things we perceive is what scientific observation and seeing that are all about. We may see something and wonder exactly what it is we see. As is to be expected by now, such an answer requires theory, and we will get different answers depending on our theoretical beliefs. But there is nothing wrong with the view that scientific measurement, that is, gaining information about a system on the basis of the behaviour of our instruments, is theory-laden. We all see$_1$ the dial or instrumental readings. What they indicate, ascertaining what *information* they provide us about a system, is entirely another matter, one subject to theory.

What this means is that scientific observation, which is often expressed in the form of *seeing that*, is not strictly perception *per se*; what seeing that really expresses is the *acquisition of information or knowledge* strictly on the basis of the individual's visual experience or what he sees$_1$. Hanson's mistake, then, was to confuse seeing with knowledge based strictly on visual means. In order to make this clear, notice first of all that not all knowledge is based on visual means. Scientists may base their claim to know the existence of a particle, whose presence they are not observing, on other information available to them in the same way that we know that certain things happened in the past even though we ordinarily do not claim to *see that* they occurred. On the other hand, our scientists may base their claim to know on what they see$_1$: they *see that* the particle is present in the same way that a layman can see that the water in the tea kettle is boiling or a doctor can hear that a patient's heart is not normal. A careful analysis of the journey from see$_1$ to *seeing that* is precisely where Hanson's analysis of scientific observation is amiss and why his formulation of perceptual relativity in the sciences falls short of the mark.

Let us now turn to Dretske's analysis of *seeing that*. *Seeing that* or scientific observation claims involve the ascription of properties to things. One may see that the banana is ripe or that a person is getting angry. So we will give these claims the general form, '*S sees that b is P*' (where '*S*' refers to the observer, '*b*' the object or state of affairs observed and '*P*' what is predicted of *b*). What we need is a set of conditions or criteria that will take us from seeing$_1$ *b* to *seeing that b is P*.[27]

If the reader will recall, since it is a conceptual fact that one can neither see$_1$ what is not the case, nor see that which is not the case, seeing that *P* entails that, as a matter of fact, *b is P*. So, our first condition for the truth of *S sees that b is P is simply*:

(i) *b* is *P*.

Of course, seeing that *b* is *P* is based on what *S* sees. Even though Dretske allows for cases of *S* seeing that *b* is *P* where *S* does not actually see$_1$ *b*, he nevertheless argues convincingly that these cases are parasitic on a notion of seeing that *b* is *P* on the basis of seeing$_1$ *b*. Normally, when *S* sees that *b* is *P*:

(ii) *S* sees$_1$ *b*.

So far, things should be fairly obvious and relatively non-controversial. The third condition takes into account the above-mentioned connection between *seeing that* and the visual aspects of the situation confronting the observer. Any scientific theory worth its salt must include statements about how its subject matter characteristically behaves, looks or appears the way it does in relation to its properties. For example, optical theory may tell us that red things will look purple when illuminated under a blue light or a psychological theory may claim that a form of compulsive behaviour is a manifestation of a particular type of psychosis. Given that such claims are usually qualified by the conditions under which *b* is observed, we have

(iii) The conditions under which *S* sees$_1$ *b* are such that *b* would not look, *L*, the way it does unless it was *P*.

Condition (iii), then, expresses something of paramount importance about *seeing that*: if things did not have a characteristic look, L, under specified conditions, we could not know or detect their properties by visual means (although we may come to learn them by other means, either through other senses or, indirectly, by the way they affect things we can see). Again, the reason why Kepler is unable to see that the Earth is turning into the sun's rays is that there is no special way such motion appears from his vantage point in space; nor does it possess a look that distinguishes it from the sun moving around the Earth.

Even if (i)–(iii) hold, it obviously does not follow that 'S sees that b is P' is true; after all, S may be ignorant of the fact that b would not look, L, the way it does unless it was P. While ignorance does not affect one's ability to see$_1$ b, it surely can put a halt to seeing that b is P. In order to complete the set of conditions for the truth of 'S sees that b is P', then S must possess information about how b looks in relation to b's characteristic features, i.e.,

(iv) S, believing (iii), takes b to be P.

If S did not have a theory about how bs in general look in relation to being P then S could not know that b is P by visual means, i.e., S could not see that b is P. This is how theory and perception are related: theory tells us *what* we are perceiving; it identifies what we are measuring.

It is implicit that (i)–(iv) hold for any form of scientific observation, not just seeing that. Let me run through some examples. Suppose two philosophers have this conversation while in the process of making tea:

Philo 1: Is the water boiling yet?
Philo 2: Yes, it is.
Philo 1: How do you know?
Philo 2: I can see that it's boiling.
Philo 1: What do you mean 'You can see that it's boiling'?
Philo 2: I can see the steam escaping from the kettle, and teakettles just don't look this way in this situation unless the water inside is boiling. Q.E.D.

Or, they might have another conversation:

Philo 1: Is the water boiling yet?

Philo 2: No, but it's about to boil.
Philo 1: How is that?
Philo 2: I can hear that it's about to boil.
Philo 1: How can you do that?
Philo 2: The kettle is making this vibrating noise and it wouldn't sound that way unless the water inside was about to boil. Hey, it's boiling!

A scientist can even smell that something is the case under the right circumstances. For example, any chemist can smell that hydrogen sulphide is being used in an experiment because it smells just like rotten eggs. And (i)–(iv) are especially designed to handle – as they must – observations made by means of scientific instruments. This extrapolation to instruments requires a slight modification of (iii) and (iv) but the idea is basically the same.[28] Electromagnetic theory tells us how a galvanometer will look or behave when an electric current is passing through it. We can thus use the deflection of an attached mirror to measure or observe the presence and amount of an electric current. This instrumental variation on *seeing that* requires that we see_1 the behaviour of the instrument and possess a theory which links the behaviour of the instrument with various types of things that happen to it. In the galvanometer case, we say: (iii) conditions are such that the galvanometer's mirror would not be deflected (in the way it is) unless there was a specific amount of electric current passing through the circuit; and (iv) we, believing conditions as described in (iii), take it that a specific amount of current is passing through the circuit.

Admittedly, my presentation of Dretske's analysis of scientific observation is cursory, leaving out its many intricacies, qualifications and restrictions. But the major objective of presenting his analysis to the reader is to show that Hanson's gestalt approach does not do justice to the complexities involved in establishing or identifying *what* the scientist observes given the observing situation and the theories at work. Another objective is to make precise the notion of perceptual relativity in the sciences, and I feel that Dretske's (i)–(iv) have done this to a great extent. *Seeing that* is relativised, via (iii) and (iv), to scientific theories and beliefs concerning how things look, appear or behave in relation to what they are. As these theories change, it is likely that how we appreciate what we perceive will change as well.[29]

Finally, if we want to take the claim that theory determines what we can observe beyond that of being a philosophical cliché, the exact nature of the relationship between theory and observation must be revealed in order to understand how it is that the former determines the latter. Dretske's analysis has provided us with a very important clue on how to proceed.

NOTES AND REFERENCES

1. I. Kant, *Critique of Pure Reason*, trans. N. K. Smith (London: Macmillan, 1958) pp. 129–57.
2. Wittgenstein, *Philosophical Investigations* (Oxford: Basil Blackwell, 1958) pp. 193–215.
3. Hanson, *Patterns of Discovery* (London: Cambridge University Press, 1958) pp. 13–14.
4. Ibid., p. 31.
5. Ibid., pp. 19–20.
6. Ibid., p. 2. See also Hanson's, *Perception and Discovery* (San Francisco: Freeman, Cooper, 1969) pp. 77–90.
7. 'Understanding how phenomena are sometimes felt to become comprehensible when viewed through a particular theory is somewhat analogous to the "Gestalt click," and this is itself instantiated *via* the pictorial illustrations which support the analogy I have been shaping', in *What I do not Believe and Other Essays* (Dordrecht: Reidel, 1972) pp. 28–9.
8. Hempel, *Philosophy of Natural Science*, pp. 40–6.
9. Hanson, *Patterns of Discovery*, pp. 60–2.
10. Ibid., p. 57.
11. Ibid., pp. 60–1.
12. Ibid., p. 62.
13. This is argued for throughout Hanson's *The Concept of the Positron* (London: Cambridge University Press, 1963).
14. Ibid., pp. 25–41.
15. Hanson, *Patterns of Discovery*, p. 85. Most of this material is in Chapter IV of Hanson's *Patterns of Discovery*. See also pp. 288–300 of *What I do not Believe and Other Essays*.
16. Hanson, *Patterns of Discovery*, p. 81.
17. Ibid., p. 86.
18. Hanson, *Observation and Explanation* (New York: Harper & Row, 1971) p. 64.
19. J. C. Smart, *Between Science and Philosophy* (New York: Random House, 1968) pp. 75–6.
20. Hanson, *Patterns of Discovery*, p. 61.
21. Hanson, *The Concept of the Positron*, p. 44.
22. Hanson, *Patterns of Discovery*, pp. 11, 15.
23. This distinction occurs in F. Dretske's *Seeing and Knowing* (University of Chicago Press, 1969), pp. 37–8.

24. Galileo, *Dialogue Concerning the Two Chief World Systems*, trans. S. Drake (Berkeley: University of California Press, 1970) pp. 126ff.
25. Dretske, *Seeing and Knowing*, p. 20.
26. Ibid., p. 17.
27. What follows is an encapsulation of Chapter III of Dretske's *Seeing and Knowing*.
28. This is worked out in detail in Chapter IV of Dretske's *Seeing and Knowing*.
29. How this takes place is rigorously described in Chapter V of Dretske's *Seeing and Knowing*.

5 Science as a human activity: Kuhn's concept of a paradigm

In Chapter 3, I briefly discussed the dispute between the formalist, contextualist and realist views on the nature of scientific theory. There is little doubt that Kuhn's approach to science is the penultimate contextualism, one diametrically opposed to the formalists, especially to the adherents of D–N. Kuhn also disavows the doctrine of scientific realism because it supposedly can not provide us with a tenable model of scientific change (or progress). Instead of viewing the transition from one fundamental scientific theory to a new, 'better' system in terms of what many of us would normally think of as 'coming closer to the truth' or 'more accurately describing the world *out there*', the process of scientific change is actually based on non-objective, sociological-political factors.[1] The notion of scientific transition being an objective process is a myth that has been perpetuated by the logical positivists, formalists and realists, and the sooner it is debunked, the better. Scientific progress must be viewed not in logical but human terms, in particular, in terms of the replacement of a total network of values, commitments and theoretical institutions of a scientific community with an entirely different set. Scientific change is more like an evolutionary struggle of ideas where the persuasive – *vis-à-vis* logical – force of a theory is a measure of its survival traits.[2] Hence the structure of scientific progress mirrors that of a revolution where logic and objectivity of the D–N kind play little or no role in the process. Such a result actually follows from Kuhn's analysis of the nature of scientific theory, in particular, his development and application of the notion of a scientific *paradigm*, to which I now turn.

1. WITTGENSTEIN ON MEANING AND USE

Instead of going directly to Kuhn's treatment of the concept of a scientific paradigm, it might be pedagogically more useful to examine the concept's origin, which can be found in Wittgenstein's critique of the formalist approach to language. The ideal language philosophers felt that logical syntax captured the essence of language, supplying us with this recipe for determining the meaning of an expression: translate the expression into logical form and then identify its referents or extensionality. This is the formula that meaning equals logical syntax plus reference mentioned in Chapter 3, something Wittgenstein forcefully disputed in his *Philosophical Investigations*.

Let us outline Wittgenstein's critique of the above approach to meaning and what he felt should be its replacement. According to ideal language philosophy, all descriptive language has logical syntax as its common 'grammar' while the unit of meaning is the proposition or sentence. To the extent a sentence or expression fails to meet the standards set by symbolic logic it is vague, requiring analysis. Language is essentially an interpreted calculus which is often besmirched by emotional, imprecise, non-objective, anthropomorphic elements in ordinary speech. Ordinary language is notoriously vague. According to Wittgenstein, the above view completely misconstrues what language, namely a system of communication, is all about. Contrary to the formalist approach, translating ordinary language into the symbolic mode does *not* serve to clarify the meaning of an expression; on the contrary, part of the message is lost in the translation, precisely because the above so-called vague, non-referential and non-logical elements are left out. If we truly want to understand what is meant by a particular linguistic expression, we can not possibly avoid going into things such as the intent, motivation and basic needs of the speaker. These are all non-logical features of language.

For language is a social activity among many other human activities.[3] We communicate by performing deeds, i.e., we do not merely write down symbols or make noises but do things with words, symbols, etc. Any good theory of meaning must capture what it is we are actually doing by uttering 'P', writing 'P' down on a sheet of paper, and so on.[4] Logical syntax really can not reflect this aspect of language, for while getting at the point of what is being said is partly a linguistic question, it is only the tip of the iceberg in that any thorough analysis of what is meant must ultimately bring in the *non-linguistic* attitudes of the speaker, the social context in which the utterance occurs, the values of the speaker and listener, their goals and commitments, and many other things implicitly underlying the act of communication.[5] Wittgenstein introduces the notion of a *language game* to denote the social context in which the expression occurs. The unit of meaning, then, is no longer the proposition or sentence but it is *relativised* to a language game: instead of the meaning of an expression being a function of its logical syntax plus reference, *its meaning is a function of its use within a language game*.[6]

In order to illustrate the contextual complexities underlying the use of an expression, consider the concept of 'King' in chess. We can teach a child to point to the King, move the King under a variety of circumstances but does it follow from this that the child has the concept, King? Very unlikely, for the child has no familiarity with the rules of chess, board games, making a move in a game, strategies, winning or losing, realising that the King is the most important piece on the board, etc. All of these things and much more underlie our conceptual repertoire. Naming the King is just one minute aspect of it.[7] Wittgenstein calls these background features underlying the use of a concept, *forms of life*. Having the concept in question involves skilfully participating in its respective forms of life. They 'breathe meaning' into what we say.[8]

Even the simplest, most 'primitive' forms of communication presuppose a highly complex system of socio-psychological needs and commitments shared by the participants. Imagine two members of a primitive tribe who have been assigned the task of constructing a stone hut.[9] The artisan and his helper have a very limited vocabulary of two words: 'slab' and 'brick'. When the artisan says 'slab!' his helper goes to the slab pile, picks up a slab, and carries it over to the hut. When 'brick!' is uttered, a brick is carried to the hut. Now even if 'slab!' were elliptical for 'Bring the slab over here' or some other logically complex sentence, something Wittgenstein considers to be highly artificial and unlikely, what difference does it make in terms of capturing what is meant in this social context?[10] Explicating the meaning here amounts to revealing, among other things, an implicit cooperation between the artisan and his helper, tacit agreements between the two builders and their tribe, their intent to bring something about by the use of linguistic expressions, and so on. Without this information, how could we determine the point of 'slab!' and 'brick!', i.e., their meaning?

Not only are language games devices that serve to help us get at the meaning of what is being said, they enable us to compare the meaning of expressions as we go from one social context (form of life) to another.[11] If we want to compare descriptive discourse to, say, ethical discourse or even discussions on the relative merits of works of art, we must above all examine the corresponding language games in which these acts of communication take place. Language games, then, serve as means of comparison or as *paradigms* by means of which we are to gain insight into what is meant by a particular linguistic expression. In so far as there is no single language game or social context which is common to a great variety of linguistic activities, ranging from very simple acts of communication such as the 'slab'–'brick' language game to the extremely complex language game of ethics, there is no real essence of language;[12] so, even if it could be shown that all these language games or systems of communication have a common logical syntax, such a dubious achievement would have no bearing on the issue of there being an essence of language. If Wittgenstein is correct, language is indeed a far cry from that 'cold', static formal calculus of the ideal language philosophers.[13]

Let us review the above points. Language games are a technique for getting at the true meaning of expressions, replacing logical syntax plus reference. Discourse does not take place within a social vacuum; so, in a way, ascertaining the meaning of what is said helps to reveal the most fundamental features of our lives which underlie our use of words such as our commitments, beliefs, how we view the world, etc. – our forms of life. Another way to put this, for want of a better word, is to say that our activities are *rule-governed*. Part of evaluating the meaning of what is said is to bring out the rules underlying the particular language game in which it is said. Admittedly, the use of 'rule' here can be very misleading, for rules are usually conventional in nature, being either chosen or accepted by groups of individuals. The kind of 'rules' I am talking about here are more or less forced upon us by the dictates of life; there is no choice about them in that we could not cope or successfully do whatever it is we are doing in their absence.[14] They also reflect our most fundamental needs in what we do.[15] Of course, we do not justify them by appealing to logic; if so, they would cease to be fundamental or rock bottom. In a way, knowing these rules and commitments underlying language games will reveal socio-psychological facts about the way we live in the deepest sense.[16] Logical syntax plus reference simply keeps everything about *us* on the surface.

The rules underlying our use of language in a variety of activities are naturally contrasted with laws of nature. Rules can always be broken. However, in the same way that you *can* move your King eight squares across the board but *not* make a legitimate or any kind of a chess move for that matter, saying something like 'I know that I promised to pay you back but I feel no obligation to do so' or 'I know this will hurt you and you do not want me to do it but I'll do it anyway because I feel like it' reveals someone who can not deal with others ethically, someone who lacks a form of life in a way very much like the child who can pick out the King but can not in the least play chess.[17] And, in the same way that the rules and strategies of chess help us make sense of various moves by the players, we speak of rules underlying a variety of linguistic activities in order to make sense of what is being done by particular speech acts. Without them, the linguistic act could not be performed or, at least, done correctly. In other words, the sense of 'rule' I have in mind here is intended to bring out the motivational features or rock bottom commitments underlying our use of words. They often reflect things like the socio-psychological needs of an entire society or community of individuals. The ultimate semantics, then, is a combination of social psychology and value theory.[18]

2. SCIENTIFIC PARADIGMS

We now turn to Kuhn's use of 'paradigm' in analysing the nature of scientific theory and scientific change. Ever since its inception in *The Structure of Scientific Revolutions*, the concept of a scientific paradigm and its application to various problems in the philosophy of science has undergone a wide range of attacks from several philosophers of science, the major criticism being that such an important concept lacks sufficient clarity to be of any use. The concept of a paradigm, they say, is used to handle too many unrelated tasks and, as a result, Kuhn fails to supply us with a core meaning.[19] In spite of all this controversy, I feel that since this concept clusters around so many of the above Wittgensteinian ideas on language, it would be much better to keep within such a framework if it leads to a clarification which answers at least some of Kuhn's critics.

What is a scientific paradigm? According to my view, a scientific paradigm is a network of rules and commitments that make sense of a variety of interrelated scientific activities. And, in the same way that the rules underlying language games served as a means of comparing different forms of discourse, we compare different scientific theories not in terms of their outward formal structures but rather in terms of their respective networks of rules and commitments. In other words, science is just like any other human activity, subject to its own set of rules and forms of life. This is the penultimate contextualism: any correct analysis of a theory must include those *normative* features underlying a scientific activity;[20] different theories reflect different networks of rules. (I realise that my presentation of Kuhn's central concept actually contradicts one of his major contentions in *The Structure of Scientific Revolutions*, namely, his claim in Chapter 5 that there are no rules underlying scientific behaviour. I will discuss this below but I feel that presenting the notion of a scientific paradigm in terms of a system of rules actually does far more justice to the notion than Kuhn realises. So, even though Kuhn's use of 'paradigm' departs somewhat from Wittgenstein, I argue below that such a departure is a mistake.)

Among the scientific activities of which paradigms as networks of rules and commitments are to make sense: designing experiments, making calculations and predictions, interpretation of data, applying the formalism of a theory to specific phenomena, supplying explanations, 'puzzle-solving' or assimilating seemingly disparate phenomena into the theory, and many others. These everyday activities of the scientific community go under the Kuhnian heading of *normal science*. The network of rules and commitments around which these activities .cluster serves to identify the fundamental theory held by that particular scientific community. Explanation, for example, is an activity or technique which shows the scientist and his peers how data are assimilated or processed in accordance with their paradigm. Data that can not be so handled by the paradigm resist comprehension and our

scientists must find a technique which somehow fits the recalci-
trant data into the overall scheme of things.[21] Thus, unlike
Hempel, explanation is not atomic but, as with Hanson, it brings
to bear an entire system of beliefs, commitments and even values
of a scientific community.

Notice that a scientific paradigm is not simply a set of rules and
commitments of a scientific community but it is an ordered set,
i.e., the elements of the paradigm are interrelated or structured in
such a way to form what Kuhn calls a *disciplinary matrix*. The
matrix has four basic elements: (1) the symbolic generalisations
(laws of nature) and definitions of D–N; (2) metaphysical models
or ontological commitments of the scientific community; (3)
values or norms of accuracy, error, etc.; and (4) exemplars or
shared examples, that element of the matrix which is supposed to
best capture Kuhn's original use of 'paradigm'. (1)–(4) are
intended to make sense of a wide range of scientific activities or
behaviour.[22] They are a basis for the sociology of science.

What Kuhn has in mind by (1) is fairly straightforward. He is
simply acknowledging that there is a formal aspect to almost any
scientific theory; but even though any theory possesses a system of
formal representation, there of course is much more to theory than
a set of laws and conditions in the same way that meaning goes
beyond logical syntax plus reference.

Metaphysical models more or less comprise the scientific
community's fundamental beliefs about the nature of things and
how things work. Recall that the logical positivist would most
emphatically deny that a scientific theory meaningfully contained
anything like (2). But Kuhn insists that such beliefs play an
essential role in the very functioning of a science; (2) is especially
at work when it comes to scientific observation. Like Hanson,
observation is theory-laden and our metaphysical models, in a
very important sense, determine what the scientist can observe.
Such models might reflect a community's commitment to the
corpuscular view of nature or the contrary belief that individual
things are ultimately properties of fields. Whether the scientist
views nature in terms of atomism or as a universal field will affect
how data are interpreted and explained. For example, compare
the radical difference in their respective treatment of motion. The
former speaks of the change of location in time of one and the same
physical object (a configuration of atoms) while the latter views
motion as a disturbance, *vis-à-vis* an object, moving through

space, where a disturbance is not a substantial object but a property of a field. (The reader should be familiar with the doctrine of theory-ladenness by now.) This holds true of the psychological and social sciences as well. A psychologist of a behaviourist bent, who views human beings as complicated reflex systems is, indeed, working under a different paradigm than, say, the Freudians and cognitive psychologists. This is just one of the many ways that an element of a disciplinary matrix can affect the others.

Even our standards of accuracy in measurement and confirmation are relativised to a paradigm. More important, these standards are inextricably tied to the values and norms of the scientific community, i.e., it is a myth that measurement and theory-testing are completely objective, value-free processes as the logical positivists and formalists would have us believe. Take a term used to express degrees of accuracy like 'exact'.[23] Suppose a friend invites you to dinner and you ask what time you should arrive. Your friend answers that quiche lorraine will be served and you should show up *exactly* at 6:00 p.m. so the serving will be neither too hot nor too cold. When you arrive for dinner, the church clock across the street is striking six. However, upon entering, your friend in a piqued tone of voice says that you are late, pointing to an atomic clock on the table. What this ridiculous example shows is that words like 'exactly', 'precisely', 'correctly', 'approximately', etc. are relativised to things like our goals, intentions and our conception of things underlying our experiments. What may be a perfectly good measuring device for one situation might be ludicrous when used in another. For example, imagine using a stopwatch to time the journey of a particle in a linear accelerator. Thus, Kuhn contends, even experimental procedures are value-laden as an integral part of a scientific paradigm.

Before going on to (4), there is something that must be pointed out about the above claim that measuring and experimental procedures are value-laden. It may be the case that Kuhn's claim rests upon an equivocation. For it is one thing to claim that there exists a set of rules or guides for what counts as an acceptable experimental procedure and, because they do serve as standards which tell the scientist if things are being done in a correct manner or not, *they* are normative or value-laden. But it is quite another thing to claim that our standards of accuracy are value-laden in the sense that they depend on some *other* values that the community of scientists may happen to have. While the former claim is not a very controversial one, the latter is highly contentious. For example, some philosophers of science have argued that the more socially obnoxious a theory is, the tougher our standards of confirmation turn out to be. If confirmation of our theories is like this, then we can say farewell to objectivity in science.[24] I will return to this in the next chapter.

It is not very clear what Kuhn means by 'exemplars' or 'shared examples', and things seem to get even more confused in his subsequent attempts at clarification. My guess is that exemplars are devices that somehow serve to identify the nature of a phenomenon that first appears to be entirely unlike anything that has confronted the scientist before.[25] Unless the scientist can relate this phenomenon to other, known phenomena, he is in no position to apply the theory's formalism in this new situation. For example, a student who is at home with applying Newton's laws to collision phenomena and the motions of the planets may be unable to handle the motion of a pendulum. This difficulty is surmounted only when the student recognises that the behaviour of a pendulum is actually just another case of gravitational phenomena.

> The student discovers, with or without the assistance of his instructor, a way to see his problem as *like* a problem he has already encountered. Having seen the resemblance, grasped the analogy between the two or more distinct problems, he can interrelate symbols and attach them to nature in the ways that have proved effective before.[26]

The problems with which the student is already familiar and which are used as a basis for extrapolation appear to be exemplars or shared examples. How is the student able to use exemplars to extend the formalism of a theory to an entirely new situation? Kuhn starts out by mentioning rules but later denies that this is the case. He finally settles upon a gestalt approach, one very similar to Hanson's. The student simply 'sees' or 'gestalts' that the two distinct phenomena are sufficiently similar to warrant the application of the same set of equations.[27] Unfortunately, I find Kuhn's account not much more intellectually satisfying than saying that the student simply realises that the two phenomena are quite similar after all. The mystery still remains: how is it that a scientist can come to conclude that two phenomena which appear to be radically different in nature are actually the same type of phenomenon? By appealing to a gestalt model, Kuhn actually missed a golden opportunity to make excellent use of (2) in his disciplinary matrix, for in Chapter 10, I shall argue that such a conclusion is not the result of some sort of gestalt but it is carried out instead by what I will call providing a common ontology or metaphysical model among the differing phenomena. The details of this will be worked out in that chapter.

I said earlier that (1)–(4) form a disciplinary matrix. How are they related within the matrix? Kuhn never explicitly says. However, we do know that a change in any one of the elements may have a significant effect on all the others since they are all interrelated in a delicate and complicated way. (1)–(4) make sense of a variety of scientific activities. Like the rules underlying a language game, they tell us what would *count* as a prediction, observation, experiment or even a fact. They determine what the relevant data are, what questions we should ask, what experiments to perform, how to handle new data, what counts as evidence for or against a supposition, what counts as a good or bad experimental result, and so on. But this means that the values of the scientist and commitments of the scientific community can no more be separated from the identity of a fundamental scientific theory than the meaning of an expression can be stated independently of the social context (commitments and forms of life) underlying a language game. Scientific theory and practice, fact and value are not separable. The result of this is relativism in the sciences of a most fundamental kind.[28]

3. THE STRUCTURE OF SCIENTIFIC REVOLUTIONS

I have characterised a scientific paradigm as a device that makes sense of a variety of scientific activities by showing how they are in accordance with a network of rules and commitments of the scientific community. However, not all scientific activity falls under this rubric, for under certain rare circumstances, the scientific community finds itself in a situation where the participants of the paradigm are forced to ask questions about the rules and working commitments themselves. This is not normal science but *meta*-science, namely those activities of the scientific community which concern the acceptability of their paradigm, whether this particular 'game of science' should be played according to these rules. One of Kuhn's major claims seems to be this: while we can speak of a scientific judgement or practice being correct or incorrect, rational or irrational, objective or non-objective, etc., within the framework provided by the paradigm, these categories of assessment break down when they are applied *across* paradigms. There are no rules which allow the members of a scientific community objectively to evaluate fundamental scientific theories, no more than we can objectively say that soccer is a better game than chess or that a play in soccer is better than a move in chess. Contrary to what we have been taught by popular accounts, scientific change or scientific progress is not an objective, continuous process but one which is driven by social-political forces that lead to the eventual upheaval of the old scientific conceptual framework and its replacement by an entirely new set of scientific institutions. Scientific progress can not be characterised as a linear accumulation of knowledge.[29]

Let us carefully examine Kuhn's reasons for this startling hypothesis about scientific progress. In the first place, the view of theory confirmation and refutation put forth by the logical positivists and formalists is incredibly naive when it is compared to actual scientific practice in the past and present. Confirmation of a hypothesis is supposed to come about by generating predictions and then 'looking' to see if the predictions are true. If they are true, the hypothesis is confirmed; if not, it is refuted. It is as simple as that. While such a picture of theory acceptance may appear to be logical, even commonsensical, the real scientific world does not work that way at all, and there are several case studies in the history of science that bear this out.

Most of the above-mentioned studies can be found in Kuhn's *The Copernican Revolution*.[30] Even though the heliocentric hypothesis led to better predictions of the (apparent) motions of heavenly bodies, the scientific community still adhered to the geocentric astronomy. For one thing, as Popper has constantly stressed, logic dictates that it would take an infinite number of positive results to confirm an hypothesis. We can not logically infer the truth of a hypothesis from the truth of its consequents without committing the fallacy of 'affirming the consequent'. (The details of this are to be discussed in the next chapter.) So even if the heliocentric view leads to positive results, that, by itself, does not suffice to confirm it, just from logical considerations alone. The refutation of a theory is not subject to logical considerations, as evidenced by the fact that several hypotheses have been maintained in spite of negative observation results. *Modus Tollens* does tell us that a single negative result *should* falsify a hypothesis. But, as we have seen, theories are more than propositions and their logical consequences; there are other, more crucial considerations at stake here when it comes to the rejection of an hypothesis. It is not simply a matter of preserving truth as we go from the theory to its logical consequences but *whether science can function at all once the theory is purged*. Very often, the scientific community is not willing to pay the price of rejection by false predictions; and for good reasons! At the time of its inception, the heliocentric view was not incorporated into a tightly woven conceptual framework of physics, biology, theology, etc., in the way that geocentric astronomy was. Removing the geocentric view of the universe would cause the whole system to unravel. For example, we could not explain why bodies on Earth fall in terms of their natural tendency or potential to move towards and eventually rest at the centre of the universe. The heliocentric view did not supply the scientific community with a replacement for Aristotelian physics and the teleological view of the universe. Another example can be found in the successor to Aristotelian physics. Newtonian physics specifically predicts that the diameter of the elliptical orbit of a planet will not shift with time; yet, the astronomers accurately observed such a shift in Mercury's orbit. The observation would not go away, but this did not shake the confidence of the Newtonians who immediately set out searching for the planet Vulcan. We all know that the overthrow of Newtonian mechanics would be no mean achievement, that it would take centuries to find an acceptable paradigm to replace it.

How does such a paradigm swap take place if not by falsification? Kuhn begins by introducing the notion of a scientific *anomaly*. Anomalous data do not simply falsify the predictions of the theory but they resist *any* form of conceptual assimilation and because of this, they strike at the very heart of the working paradigm. They are like miracles in that this type of a phenomenon tends to leave the scientist speechless. It is an interesting feature about our conceptual scheme that it has all kinds of built-in devices which enable us to handle experiences that do not fit in with our normal expectations. For example, if we had strong reason to believe that there was a biological connection between the species, crow, and their being black, if we were to happen upon a white-feathered crow, we would nevertheless preserve our belief by calling our specimen a mutant crow; or, a substance that behaves exactly like iron except for a few properties is an isotope of iron; mirages are used to explain why bodies of water and islands are not located where we expect them to be, and so on. There are many other concepts we could add to our list such as 'illusion', 'hallucination', 'artificial', 'adulterated', 'fake', etc., that enable us to explain away those experiences that do not meet our expectations. Without them, we would be left speechless; we simply would not know what to say.[31] The notion of an anomaly is linked up with these conceptual devices. Anomalies are not phenomena that show we were wrong so much as they put us at odds end to describe what actually happened. It is not at all in our conceptual repertoire to say what took place. Suppose while you are writing at a desk, you see your coffee mug turn into a live hamster and you also know that there are no magicians of an impish character in the neighbourhood.[32] What ordinary beliefs or theories could such an event possibly falsify without a complete revision of our biological and physical conceptions? What if this occurrence happened again a year later and in the presence of others? Scientific anomalies are just like the mug–hamster transmogrification case. The scientific community is confronted with a persistent phenomenon that can not be explained away or assimilated by conceptual devices which are provided by their theories. If the anomaly is handled at all, it is accomplished by going *outside* of the paradigm. For example, in the mug-to-hamster case, one might appeal to an act of God. In the same way, the proponents of the geocentric view went beyond to invent epicycles in order to explain away the disparity between their theory and the observed motions of heavenly bodies. This can always be done – but at a price, like any other *ad hoc* move. For this reason, appealing to conceptual devices beyond our presently working system of beliefs in order to handle infelicitous data leaves us uneasy, especially when the phenomenon refuses to go away. Such persistence can only urge us to expand our 'limited' conceptual repertoire in order that incorporation be achieved.

All of this sets off a struggle within the scientific community.[33] There are the defenders of the old ways who pretend things are as usual in face of persistent anomalies, hoping that their paradigm will eventually find a way to solve these mysteries while this 'old guard' find itself confronted with a newly formed group of 'revolutionary' scientists who seek to replace the old, tottering paradigm with one more representative of their generation. After a buzz of activity, which may go on for generations, except for a few diehards, the old paradigm is abandoned by the scientific community in favour of a new set of practices, commitments and values, one which may not in the least resemble its predecessor. The revolution has been completed and there is no turning back.

The outcome of Kuhn's model of scientific change is that our commonsensical view of science as an activity that luxuriates independently of our frail humanity is a myth. For the logical positivists, the formalists and the realists, science was the standard-bearer of objectivity. Scientific change or progress as the replacement of a theory by a 'better' one was thought to be a glorious example of this objectivity. Now we find that objectivity does not preside over change; instead scientific progress is brought about by social-political forces. There is a role for objectivity in science, however, but it is a restricted one: objectivity or the application of objective standards is an *internal* consideration, one relativised to the paradigm in question.[34] Whether a given set of values and commitments embodied by the paradigm is the right set of rules for doing science is not really a legitimate question, for it would be more like asking if we should accept this way of doing things. This is not a question of truth or fact but one of value.

In the *Tractatus Logico-Philosophicus*, Wittgenstein wrote this during an earlier period of his philosophical development:

> The right method of philosophy would be this. To say nothing except what can be said, i.e., the propositions of natural science, i.e. something which has nothing to do with philosophy.[35]

As we have seen, Wittgenstein changed his ways later and with Kuhn, it is now the other way around: what natural science really says has everything to do with philosophy.

4. THE INCOMMENSURABILITY OF FUNDAMENTAL THEORIES

In order to see how Kuhn's treatment of scientific change radically departs from philosophical tradition, let us take a closer look at the approach to scientific progress he so emphatically rejects. The thesis that scientific change is a linear, continuous process that brings about the accumulation of knowledge is often referred to as the approximation view of science: scientific progress is a process of approximation, i.e., of asymptotically getting better, more accurate descriptions of nature.[36] Let 'θ_n' denote the old theory and 'θ_{n+1}' its replacement. The approximation view holds that θ_{n+1} does a better job of capturing nature as it is; although there may never be a theory in the history of cognition that completely and perfectly describes nature to every last possible detail, we can nevertheless speak of making progress in terms of approaching such a final description. Of course, such a view of progress really amounts to saying that θ_{n+1} is better than θ_n in that θ_{n+1} comes closer to the truth. Not that θ_n is entirely false; on the contrary, the approximation view suggests that θ_n, being a rougher description of nature than θ_{n+1}, can be thought to capture nature to some extent if its range is properly restricted. I will call this the *correspondence principle*: as we go from θ_n to θ_{n+1}, when θ_{n+1} is taken to some limit corresponding to the objects of θ_n, the laws and descriptions of θ_{n+1} become identical to those of θ_n, i.e., θ_n is true to the extent it can approximate θ_{n+1}'s description of nature. For example, even though we know that Aristotle's claim that the velocity of a moving object is proportional to the applied force is false, the correct law being that acceleration is proportional to force, such a description is perfectly fine when it comes to the motion of bodies in a fluid (Stokes law) *vis-à-vis* a vacuum. We also know that as the velocities of bodies get smaller and smaller compared to that of light the laws of the special theory of relativity approach classical mechanics; and, as the quantum numbers of microparticles approach infinity – 'in the large', so to speak – the quantum mechanical Schrödinger equation becomes the classical Hamiltonian. Likewise, no matter how deep and complex our psychological theories may turn out, in some sense, we expect them to fit in with our commonsensical psychological beliefs about a person's behaviour in a variety of ordinary or normal situations. So, even though θ_{n+1} is more accurate than θ_n, the latter deserves a venerable status when viewed in a proper perspective, according to the approximation view of science.

Now Kuhn feels that all of this is fatuous. It is as if we compare theories of nature in the same way we would compare the reproductive quality of a series of photographs which were taken of the same setting by a variety of cameras, each one using a different kind of film. In the photographic case, we simply choose the best likeness by directly comparing each photograph with the actual item 'out there'. But unlike the natural setting for the photographs, 'nature likes to hide'. We simply can not uncover nature as it is in order to compare our theories. Comparisons are limited to the features among our theories, certainly not between theories and objective reality.[37]

If truth is not the basis of comparison, then what is? Kuhn wants us to conclude that 'debates over theory-choice cannot be cast in a form that fully resembles logical or mathematical proof'. Rather, 'debate is about premises, and its recourse is to persuasion as a prelude to the possibility of proof'.[38] As I noted in the last section, the debate over theory acceptance is ultimately about values and it can not be settled either by proof or by experiment. One assumption of the traditional view of scientific progress was the proposition that experiments can be designed to test between fundamental, rival theories, experiments whose results serve simultaneously to confirm one of the rival theorems and refute the other. Examples of such experiments and their logic will be supplied in the next chapter. But, for now, let us note that Kuhn denies the possibilities of this kind of an experiment because the rival theories do not have enough in common to allow for their comparison.

I mentioned earlier that this was penultimate relativism. Yet, strangely enough, Kuhn denies that his analysis of theory and theory comparison leads to relativism in spite of the fact that:

> The superiority of one theory to another cannot be proved in the debate. Instead, I have insisted, each party must try, by persuasion, to convert the other.[39]

A while later, he says that theories can very well be compared in terms of 'accuracy of prediction, particularly of quantitative prediction; the balance between esoteric and everyday subject matter; and the number of different problems solved'.[40] When we add that θ_{n+1} leads to better inventions and has better puzzle-solving features than θ_n, Kuhn reassures us that his analysis does not lead us down the path to relativism. 'That is not a relativist's position, and it displays the sense in which I am a convinced believer in scientific progress.'[41] But Kuhn simply can not have it both ways, for in trying to show his critics that scientific progress is compatible with his sociological approach to science, he has implicitly, I maintain, appealed to the very approximation model of progress he explicitly rejected. The fact of the matter is that Kuhn has locked himself into an extreme relativist position whether or not he is willing to admit it.

As Kuhn's critics have pointed out, if every standard of comparison is relativised to a paradigm then there appears to be little choice but relativism. Take simplicity, for example. What may be simple in one system may be complex in another. Kuhn claims that puzzles are relativised to a paradigm. What may be a puzzle in one may not be so in another. So, how can it be said that θ_{n+1} is better than θ_n because θ_{n+1} is a better puzzle-solver when θ_{n+1} and θ_n do not, can not have puzzles in common? Kuhn's very own arguments for incommensurability make him look even more like a relativist. There are three major lines of argumentation presented in favour of this thesis. First of all, since observations (or interpreted data) are relativised to a paradigm, i.e., they are theory-laden; so scientists who champion rival theories can not really appeal to an experimental result to settle their dispute without begging the question simply because each scientist is using *his* paradigm to interpret the result in his way. Secondly, since even the meanings of theoretical terms are relativised to a paradigm, there is a communication gap. Nothing can serve as a neutral language between the theories, so rival scientists can not even agree on what they are talking about. Thirdly, there appears to be nothing that serves as a rule or guide to doing science in spite of what the formalists believe.[42] It all comes down to the values of the scientific community in the end. If these points do not constitute an argument for relativism in the sciences, then what does?

What would scientific progress really amount to if we took Kuhn's incommensurability thesis seriously? Because truth, logic and experimentation would not be the primary considerations in a debate over paradigms, the discussion would take on an air not unlike art criticism. Scientific progress would be like the development of taste in an aesthetic community. Perhaps it can not be proven that Beethoven is a better composer than Tchaikovsky or that Cezanne is a better painter than Renoir; yet, the *evolution of taste* among those who are devotees of art is pretty much an unidirectional and irreversible process. We seldom run across someone who has spent hours and hours of his life concert-going, listening to recordings and tapes say 'When I started out listening to music, I really enjoyed Bach and Beethoven but after years of music enjoyment, I have found Tchaikovsky to be a much superior, deeper composer.' In other words, as the taste of the aesthetic community gets more and more sophisticated, its artistic preference understandably changes, and not in an arbitrary way. Now aesthetic communities do have their neophytes, after all, whose experiences are limited and tastes are unrefined. But they can develop into being fully-fledged aesthetes with the proper training. The proper training, here, does not mean studying proofs or demonstrations by art critics that, for instance, Beethoven is better than Tchaikovsky. Art critics do not do such things. Rather, they ask the student to experience the work with them, point out various things here and there about its non-aesthetic features, putting the student in a position to appreciate it, to view it under the proper perspective, etc. It is to be hoped that the student will see things the critic's way in the end, i.e., will be converted to the art critic's point of view. It may be that such a process of conversion is not an objective one in that if the student still prefers Tchaikovsky to Beethoven, so much for the student in the eyes of the community. He will stand alone unless, for some unforeseen circumstance, Tchaikovsky suddenly becomes a new rage.

It appears that the above view of the development of taste most adequately fits Kuhn's account of scientific progress, and in such a way that truth need not be part of the picture.[43] As the scientific community advances in knowledge, the ontological preferences of its members become more and more sophisticated, as well as the techniques they employ for making observations, predictions, etc. But better techniques of prediction and measurement are only *incidental* to the meaning of progress, here, as the development of more sophisticated theoretical taste. Dealing with a scientist today who still maintains that Newtonian physics is true (e.g., by claiming that the shift in the orbit of Mercury can be accounted for in terms of irregularities in the shape of the sun) would be very much like the aesthetic community reacting to a music critic who maintains that Tchaikovsky is the greatest composer who ever lived. Perhaps we should just shake our heads, saying 'That fellow is really odd', which means that he is not to be taken seriously by the scientific community.

The above analogy is meant to show the reader what the incommensurability thesis is really all about. The question is whether it is a good analogy or not. But before I examine this question, let us return to Kuhn's arguments for the incommensurability of theories.

One of the more controversial issues in the philosophy of science concerns the demarcation of a meaning or concept change as we go from one theory to another. Those who believe in the unity of science maintain that the meanings of fundamental concepts are preserved as we go across theoretical languages while contextualists such as Kuhn, Feyerabend, *et al.*, feel otherwise. The latter believe that if a fundamental concept occurs within a new theoretical context, that in itself suffices to change its content. Now, how are we to decide between these rival claims about the identity of meaning across theories? Does 'energy', for example, mean the same thing in Newtonian mechanics as it does in the special theory of relativity? Kuhn's answer is an emphatic 'No' because 'energy' occurs in different paradigms. But his answer assumes that occurrence of a concept in a different context entails a difference in meaning. This is far too strong, for we do not want so easily to multiply the meanings of words. The problem here is that neither side of the dispute over the preservation of meaning under theory transformation supplies us with a criterion for the synonymy of scientific terms. Kuhn does supply us with a hint, however. The meaning of a scientific term is tied up with the symbolic representation of a theory, its generalisations which 'function in part as laws but also in part as definitions of some of the symbols they deploy'.[44] And since different theories employ different generalisations, the meanings of terms change; acceptance of a new system of symbolic representation amounts to accepting a group of redefinitions. Thus, since the laws in which 'energy' occurs in Newtonian mechanics take on a new form of expression in the special theory of relativity, the meaning of 'energy' is not preserved as we go from the former to the latter. The same applies to 'mass', 'momentum', 'force', and so on. The upshot is that Newton and Einstein could not really communicate if they were ever brought together to discuss the relative merits of their theories.

I frankly find the above result incredulous; if anything, it ought to be considered a *reductio* of Kuhn's thesis rather than an argument for the incommensurability of theories. Although I can not possibly provide the reader with a criterion of the synonymy of theoretical terms, there appears to be a fairly commonsensical distinction that can be made between an actual change in the meaning of a concept as opposed to learning something completely new about the object or property to which the concept applies. For example, learning that books burn at 451°F tells us something about books we never knew or even considered before, but such information should hardly make us revise our concepts of a book.[45] (Otherwise, like Leibniz we end up claiming that everything we know about an x is packed into the meaning of 'x', which would entail that we do not have any concept of anything unless we know everything there is to know about it.) Now imagine Newton and Einstein having the following conversation about 'energy' and 'mass'.

Einstein: You know Isaac, I made a sensational discovery about energy and mass when I formulated my special theory of relativity.

Newton: What was that Albert?

Einstein: Well, I discovered an entirely new source of energy. Not only can the motion of bodies and their location in a force field be converted into energy, matter itself can be transformed into astonishing amounts of energy. Likewise, we can convert energy into matter. This is summed up beautifully by $E = mc^2$, which is a result of my special theory. Isn't that a wonderful discovery?

Newton: Frankly, I haven't the slightest idea of what you're talking about?

So the story goes, according to Kuhn and other contextualists.[46]

Why should the meaning of 'energy' change simply because the lawlike expressions in which it occurs change? If energy is a property of a system which can be converted into work, then what difference should it make that there are different *expressions* in which 'energy' occurs? ('Energy may come from many different sources, but it's all the same thing: it makes things go.') After all, that a word can occur in different sentences does not in itself entail that the word is ambiguous. Take the concept of 'mass'. Although mass is independent of velocity in classical mechanics but velocity-dependent in the special theory of relativity, the same concept of mass occurs in both theories. Mass is a physical property of a body which is responsible for its resistance to a change in its inertial motion. Special relativity and classical mechanics may differ in the laws concerning mass, but not necessarily in their concepts of mass. Another, even more dramatic example can be found when, in 1924, L. V. de Broglie proposed that all particle phenomena could be treated as wave phenomena. (The discovery that particles had wave properties was confirmed by accident when Davisson and Germer found that electrons manifested interference patterns in the same manner as light waves.) The basis for de Broglie's work was that 'energy' occurred in two entirely different equations, each of which belonged to a completely separate theory, namely $E = mc^2$ (energy equals mass times the speed of light squared) in relativity and $E = hv$ (energy equals Planck's constant times the frequency of vibration) in quantum mechanics. These two equations from disparate systems were combined to form $mc^2 = hv$. But according to the contextualists, such a move should have been outlawed, since 'energy' *must* differ in meaning.

I feel that Kuhn's argument for contextual meaning rests on a mistake, the very same mistake that Wittgenstein pointed out when he criticised the ideal language criterion of meaning being logical syntax plus reference. The logical syntax or mathematical form of a lawlike expression may change as it occurs in different theoretical systems, but it does not follow from this alone that the scientific *use* of its terms has changed. In fact, I have suggested that, at least in the above cases, the terms were not put to a different use at all. In other words, perhaps Einstein and Newton shared enough forms of life to be able to understand one another in spite of the fact that their laws were not identical in mathematical form. (By the way, must the use of two terms be circumscribed by the same set of rules if they are to be synonymous? What if they had ninety-nine rules in common but were different in only one?) And perhaps this is why the imaginary dialogue quoted earlier about energy and mass appeared so absurd. Now, one thing Kuhn can not do is argue that Newton and Einstein can not mean the same thing or share enough forms of life to do so because their theories are, after all, incommensurable. Such a ploy would beg the entire issue.

The argument for incommensurability based on the theory-ladenness of observation is equally open to question. Kuhn contends that because the interpretation of an experimental result entails theory, citing the result of an experiment as evidence for a paradigm actually begs the question since the interpretation presupposes one's paradigm. Experimental results are always relativised to the working paradigm of the scientific community. In response to this argument, Kuhn's critics have rightly pointed out an elementary mistake he makes, resulting in a *non sequitur*: even if observation is theory-laden, the theory upon which the observation depends may be logically independent of the theory or theories that are tested. So, in the case of a crucial experiment between θ_n and θ_{n+1}, the theory entailed by the deciding experimental result may have nothing to say about the truth of θ_n or θ_{n+1}. For example, scientists may design an experiment to test between two rival theories of motion where the measurements involved entail a thermodynamical theory, say, about the length of rods as a function of thermodynamic conditions, something that has nothing to do with the rival theories of motion. I will present actual cases of this in the next chapter where I discuss the logic of a crucial experiment.

Kuhn appears to be especially confused in his rules argument for incommensurability. If anything, he is guilty of an *argumentum ad ignorantiam*, concluding that because we have come up empty handed in the search for a network of rules within a science and for rules that serve as guides for the transition from one science to another, no such rules exist.[47] The formalists can readily reply that the discovery of these rules may take a lot more time and effort than Kuhn seems to allow, that Kuhn could have used the same line of argumentation to 'prove' that great formal systems such as Euclid's geometry, *Principia Mathematica* and others should never have been developed, as their respective mental activities went on for centuries before anyone could learn about the 'rules' underlying these 'games'. Kuhn's second argument concerns the relevance of such rules, even if they are found. This argument, if anything, establishes an irrelevant conclusion.

The philosopher is at liberty to substitute rules for examples and, at least in principle, he can expect to succeed in doing so. In the process, however, he will alter the nature of the knowledge possessed by the community from which his examples were drawn. What he will be doing, in effect, is to substitute one means of data processing for another.[48]

Kuhn is arguing in the above that even if it can be shown that there are rules underlying scientific behaviour, this is not what is going on when the individual applies his paradigm, say, in recognising that pendulum motion is just another instance of gravitational phenomena. According to Kuhn, rules are not at work here; the individual simply 'perceives' or 'gestalts' things in that way.[49]

Like Hanson's argument for a new logic of discovery, Kuhn is proposing a quasi-psychological thesis to rival formal approaches to science, and he leaves himself open to the criticisms raised against the gestalt approach in the last chapter. Again, the formalist can always make this reply. It appears that *any* epistemological or inferential activity can be accounted for by saying that the participant 'sees' or 'gestalts' things a certain way if we are to take Kuhn's argument seriously. Such accounts are not very enlightening. Even more important, Kuhn has entirely missed the point of formalising an activity in the first place. The rules serve to identify what is being done, to help us distinguish between a right way and a wrong way to do it, regardless of whether the individual participating can *articulate* them. That an individual is not aware of these rules when indulging in a particular practice does not imply that there are no rules underlying the practice. Not only would the formalists be mistaken about the rules, linguists such as Chomsky would be equally wrong to believe in transformational grammars because three-year-old children and most adults can hardly articulate the rules of syntax. The only relevance the above argument against rules might have is for those who maintain a nativist stance, i.e., those who maintain that the rules are built into the psychological mechanisms of the individuals who act in accordance with them; they are in some way innately 'in us'. But formalists need not maintain such a stance in order to claim that scientific activities, including giving up one theory in favour of another, are rule-governed. Kuhn has simply confused the formalist approach with nativism.

This leads into an even more serious objection to Kuhn's socio-psychological approach to science. In spite of its attractiveness, has Kuhn actually gone too far in promoting such an approach? Notice that it is one thing to explicate a set of rules underlying a variety of scientific activities and say that these rules serve in part to characterise rational scientific behaviour, i.e., they make clear what scientific objectivity is all about. But it is quite a different consideration to ask if the behaviour of the members of the scientific community is in conformity with these rules. Kuhn purports to be describing science; yet, he may be confusing these two kinds of questions. After all, scientists are frail, imperfect human beings just like the rest of us. They too can have their objectivity clouded by self-interest, social influence, greed, etc. But this does not mean that scientific objectivity is a myth; only that scientists do not behave objectively all the time. What is so surprising about that? However, using the socio-political behaviour of the scientific community in order to show that scientific objectivity is a myth would be like arguing that there are no rules of the game because the players notoriously cheat. What I am saying here is that Kuhn's sociological approach to science may have confused the language game of power, persuasion and conversion with truth-establishing procedures in science. In fact, I wonder if Kuhn's efforts have resulted, instead, in supplying us with a caricature of science. Very often, his socio-political analyses of what goes on in science do not appear to be descriptions of science *per se*; rather, they more aptly describe the academic, engineering and business worlds. The latter often resort to psychological-political tactics in order to have it their way. They deserve Kuhn's cynical treatment, not scientific systems.

There is yet one more, crucial distinction that Kuhn's stand on scientific progress overlooks. The considerations involved in presenting an analysis of what it *means* to say that θ_{n+1} is a better theory than θ_n are not the same as those that entertain the question how we *know* that θ_{n+1} is better than θ_n. According to those who maintain the approximation view of progress, the former involves a notion of truth known in philosophical circles as the correspondence theory: a proposition or sentence is true if and only if it corresponds to the facts. For example, the sentence 'Snow is white' is true if and only if snow is white. (Although it is not important here, the correspondence theory has undergone considerable development and refinement since Tarski's formulation which, I think, has been somewhat culminated in Kripke's recent work.[50] This recent work makes the above characterisation extremely naive.) In other words, truth is a relation between a linguistic entity (a sentence or proposition) and a non-linguistic entity (the world or states of affairs). To say that θ_{n+1} is better than θ_n would mean, for the most part, that θ_{n+1} 'comes closer to the facts', namely, its propositions better correspond to the facts. How we *know* that θ_{n+1} comes closer to the facts is a separate question. We should not confuse an area of semantics (namely, the nature of truth assertions) with epistemology (how we know that an assertion is true).

Now we can return to the analogy between the evolution of artistic taste and scientific progress. Are there any important ways such an analogy breaks down? There is at least one very important difference. Even Kuhn admits that we can speak of an overall increase in the *reliability* of our theories as we go from θ_n to θ_{n+1}. This increase of reliability is usually manifested in terms of more accurate measurements or instrumentation, a dramatic increase in the number of predictions, some of which are highly unexpected, and the development of new, improved technologies. Now there is no real counterpart to this in the artistic realm. Although revolutionary schools of art may come up with new techniques producing aesthetic properties such as 'depth',[51] 'balance', 'powerful composition', 'graceful lines', etc., it does *not* mean that these new ways are better or more reliable than the old; they are simply different, which, by the way, is what makes them so interesting in the first place. If this is so, however, then Kuhn can be presented with this challenge: can he really maintain a view of scientific progress as 'a unidirectional and irreversible process' without presupposing the very role he denies to the notion of truth? How else can we *explain* why our new theories give better predictions, lead to better measuring instruments such as atomic clocks, as well as to an improved overall technology except that all of these things resulted from theories that give a better description of nature, i.e., theories that are getting closer to the truth? Otherwise, why should we believe that these things are even an indication of scientific progress? One thing is certain: they can not be explained by the values and commitments of the scientific community, as if a change in the latter can automatically bring about better predictions and technology. Without the notion of truth, then, I wonder how one can believe that 'scientific development is, like biological, a unidirectional and irreversible process'.[52]

Most of the above arguments for the incommensurability of theories take place in the 'postscript' of *The Structure of Scientific Revolutions*. There is much irony here, for it can be said that Kuhn's enterprise was directed mainly against the logical positivists and the formalists. Yet, upon careful inspection, the arguments found in the 'postscript' that were intended to answer his critics relied very heavily upon principles that originated with the logical positivists and the formalists. His argument for meaning variance across theories actually presupposes that meaning is a function of the logical syntax of the sentences of a theory, while his claim that 'the notion of a match between the ontology of a theory and its "real" counterpart in nature now seems to me illusive in principle'[53] conflates questions of meaning ('What does it mean to say that θ_{n+1} is true?') and verification ('How do I know that θ_{n+1} is true?'). Finally, in the process of attacking the notion of truth serving as a standard of scientific progress, Kuhn falls back on the conventionalist view of theories that originated with logical positivists like Reichenbach.

It should be clear by now why I feel Kuhn's arguments for the incommensurability of scientific theories are a failure. However, the question still remains: can we objectively compare theories and come up with a rational basis for saying that θ_{n+1} is better than θ_n? One thing on which there is no turning back. Kuhn was absolutely right about the bankruptcy of the model of confirmation and refutation of theories we have inherited from the positivists and formalists. The failure to come up with a model that can do justice to the complexities that are found in the history of science may actually end up being the best argument for his incommensurability thesis. For this reason, I will now turn to the nature and logic of theory confirmation in order to search for such a new model.

NOTES AND REFERENCES

1. See especially pp. 198–207 of Kuhn's *The Structure of Scientific Revolutions* (University of Chicago Press, 1970) and pp. 320–39 of Kuhn's *The Essential Tension* (University of Chicago Press, 1977).

130 A Realist Philosophy of Science

2. Most of this is argued for in the 'Postscript' (pp. 174–21) of Kuhn's *The Structure of Scientific Revolutions*. The student should study this section very carefully in order to appreciate the radical but challenging conception of science and objectivity he puts forth.
3. These social features about language are emphasised throughout the *Philosophical Investigations*, especially pp. 1–85.
4. Ibid., pp. 11–13.
5. Ibid., pp. 5ff.
6. Ibid., pp. 6, 10, 13 and especially 20.
7. Ibid., pp. 17–20.
8. Ibid., pp. 8, 18. See also N. Malcolm, *Ludwig Wittgenstein: A Memoir* (London: Oxford University Press, 1958) pp. 90–3 and S. Cavell, *Must We Mean What We Say?* (Cambridge University Press, 1976) pp. 49–53.
9. Wittgenstein, *Philosophical Investigations*, pp. 3–10.
10. Ibid., p. 8.
11. Ibid., pp. 10–13; 25 and 50.
12. Ibid., pp. 31–5.
13. Ibid., pp. 38–50.
14. Ibid., pp. 37–40; 6a. See Cavell, *Must We Mean What We Say?*, pp. 9–16; 21–31; 49–52.
15. Cavell, *Must We Mean What We Say?*, pp. 26 and 38.
16. Ibid., pp. 66–9.
17. Ibid., pp. 49–52.
18. Wittgenstein, *Philosophical Investigations*, pp. 37–40; 6a.
19. For example, see D. Shapere's 'Structure of Scientific Revolutions', in the *Philosophical Review*, 73 (1964) pp. 383–94 and M. Masterman's 'The Nature of a Paradigm', in *Criticism and the Growth of Knowledge*, ed. I. Lakatos and A. Musgrave (Cambridge University Press, 1970) pp. 59–89.
20. Kuhn, *The Structure of Scientific Revolutions*, p. 207, and 'Reflections on my Critics', pp. 233, 238 and 263 in *Criticism and the Growth of Knowledge*.
21. Kuhn, *The Structure of Scientific Revolutions*, pp. 52–65.
22. Ibid., pp. 181–98.
23. Cf. Wittgenstein, *Philosophical Investigations*, pp. 41–2.
24. See especially I. Scheffler's detailed critique of this view in his *Science and Subjectivity* (Indianapolis: Bobbs-Merrill, 1967).
25. Cf. Kuhn, *The Structure of Scientific Revolutions*, pp. 187–91; 'Reflections on my Critics', pp. 273–5; and *The Essential Tension*, pp. 293–319.
26. Kuhn, *The Structure of Scientific Revolutions*, p. 189.
27. Cf. p. 73 of 'Reflections on my Critics' and pp. 305–19 of *The Essential Tension*.
28. It was this result that prompted I. Scheffler's *Science and Subjectivity*.
29. Kuhn, *The Structure of Scientific Revolutions*, pp. 198–210; 'Reflections on my Critics', pp. 259–77; and *The Essential Tension*, pp. 320–39.
30. Kuhn, *The Copernican Revolution* (Cambridge: Harvard University Press, 1959).
31. J. L. Austin strongly emphasised this feature about our language in *Philosophical Papers*, ed. J. Urmson and G. Warnock (London: Oxford University Press, 1970) throughout his 'Other Minds', and in *Sense and Sensibilia* (London: Oxford University Press, 1962).
32. One can find countless examples like this in the *Philosophical Investigations*.

Their purpose is to help us delineate the boundaries of our conceptual scheme.

33. Kuhn, *The Structure of Scientific Revolutions*, pp. 52–110.
34. Ibid., pp. 198–207 and 'Reflections on my Critics', pp. 259ff.
35. Wittgenstein, *Tractatus Logico-Philosophicus* (London: Routledge & Kegan Paul, 1922) section 6.53.
36. See K. Popper's 'Normal Science and its Dangers', pp. 56–7 in *Criticism and the Growth of Knowledge*.
37. Kuhn, *The Structure of Scientific Revolutions*, pp. 206–7; 'Reflections on my Critics', pp. 264–6.
38. Kuhn, *The Structure of Scientific Revolutions*, p. 199.
39. Ibid., p. 198.
40. Ibid., p. 206.
41. Ibid.
42. Ibid., pp. 43–51, 191–207; 'Reflections on my Critics', pp. 260ff. and *The Essential Tension*, pp. 263, 320–39.
43. Cf. Kuhn, *The Essential Tension*, pp. 340–51.
44. Kuhn, *The Structure of Scientific Revolutions*, p. 183.
45. On the other hand, suppose we were to discover that books could be eaten and, in addition to this, that the amount of nourishment a book supplied was a function of its contents. Here might be a case of learning something new about an *x* calling for a revision of our concept '*x*'.
46. This point goes back to the dispute between the positivist and the contextualist on the relationship between theory and the meanings of terms which was described in Chapter 2. The above parody dramatises how extreme the contextualist position really is. For advanced reading on this, see P. Achinstein's *Concepts of Science* (Baltimore: Johns Hopkins, 1968) pp. 91–105.
47. Kuhn, *The Essential Tension*, pp. 43–51.
48. Ibid., p. 314.
49. Ibid., pp. 306–19.
50. Kripke, 'Outline of a Theory of Truth', *Journal of Philosophy*, vol. LXXII, no. 13 (1975) pp. 690–716.
51. For example, one greatness of Cezanne was that he was able to create depth in his paintings by movement rather than by means of perspective.
52. This challenge to Kuhn's 'conventionalism' constitutes a line of argumentation for believing in the existence of theoretical entities which will be developed in Chapters 6 and 7 of the text. It essentially claims that we can not make sense of the hallmarks of scientific progress unless we affirm that our theories are getting closer to the truth.
53. Kuhn, *The Structure of Scientific Revolutions*, p. 206.

Part II

6 Confirming hypotheses

The first part of the text has been devoted to laying out what I have taken to be some of the most important approaches to the nature of scientific theory. The diversity of these approaches has led to an interesting controversy surrounding a variety of topics, such as the nature of scientific explanation, the relation between prediction and explanation, the nature of confirmation, the objectivity of scientific progress, etc. One of the more perplexing questions that has come out of these comparisons concerns the status of theoretical entities. On the one hand, we have the realist who insists that even if these entities are, in principle, unobservable the scientist has excellent reason, is rational and objective to believe that they exist because they play an indispensable role in scientific theory. They are precisely what theory is all about and, the realist maintains, science could not function to predict and explain without them. At the end of the last chapter, I suggested how the realist would go about justifying such a stance. On the other hand, there are those who wish to assign a mere pragmatic status to entities such as electrons, photons, quarks, *et al.*, in that while they may be useful for making predictions, handling data, as aids for visualising unobservable processes, etc., there is nothing in the nature of theory or the practice of science itself which dictates or entails their existence. To coin a Humean phrase, the realist's belief in the existence of entities underlying observables goes far beyond the evidence of the senses. The truth or success of a scientific theory does not entail the existence of any such things. Although many of the philosophies of science discussed in the previous chapters subscribe to this opinion to some extent, logical positivism has been the major standard-bearer of anti-realism.

Does the truth or success of a scientific theory commit one to a belief in those entities corresponding to theoretical terms or are such things just 'convenient myths' whose postulation is for pragmatic reasons alone? In this chapter and the next, I shall develop a line of argumentation which will show in what way theoretical entities play an indispensable role in scientific theory and practice. This argument for the realist position ties in an analysis or model of theory confirmation with an ontological approach to the nature of scientific explanation. It is intended to show that not only do theoretical entities play an indispensable role when it comes to a very important class of scientific explanations, theories that refer to entities underlying or beyond observables are *more open to confirmation* than theories which refer to observables alone.

The argument for realism which I will present in Chapters 6 and 7 goes something like this:

1. Rapid confirmation of a theory (or hypothesis) takes place to the extent that it can generate or assign a high probability for the occurrence of radically diverse data or test results.
2. How a theory generates or assigns probabilities to events or data is a direct function of how it explains them.
3. Theories explain, for the most part, by reducing the number of independent phenomena. Reduction takes place by identifying independent phenomena with aspects of a common, underlying ontology (= aspects of theoretical entities).
4. Realist theories account for the joint occurrences and lawlike relationships among diverse independent phenomena in a way that logical positivism, which does not allow for the provision of a common, underlying ontology, can not.

Therefore: because realism does a better job of accounting for joint occurrences of independent phenomena it is a more *confirmed* hypothesis than positivism. So, by the positivist's own standards, realism is found to be the better view.

The analysis of confirmation which leads to (1) occurs in this chapter while argumentation for (2), (3) and (4) are found throughout the next chapter. This outline is intended just as a means of reference for the reader as he progresses through Chapters 6 and 7, hoping it will give him a rough idea about my approach to defending realism.

1. CONFIRMING HYPOTHESES

Kuhn and Popper maintain that no finite number of positive experimental results can lead to the confirmation of a theory. Their position, it seems, is based in part on what appears to be a logically inviolable point: no finite set of singular propositions is sufficient to establish the truth of a universal statement. This is another way of saying that our theories are *underdetermined* by experience. We have seen that Kuhn uses the principle of underdetermination to undermine the view that the process of theory acceptance is objective in nature, and to argue that the adoption of a fundamental theory by the scientific community is actually more a case of value conversion than that of experimental reasoning. Can the principle that our theories are underdetermined carry over to Kuhn's conclusion that theories are incommensurable? I think not. On the contrary, I will attempt to show how the principle of underdetermination is perfectly compatible with the view that theories are objectively confirmed. The trick is to drop the notion of absolute or completely independent confirmation of a hypothesis and replace it with the view that a hypothesis is confirmed only relative to its rivals. The notion of comparing a hypothesis to its rivals will involve a discussion on the logic of crucial experiments.

So much controversy and paradox surrounds the topic of confirmation, I find that it would be best to limit my task here to presenting the reader with a fairly straightforward model of theory confirmation, one using principles of probability which are for the most part uncontroversial but principles that nevertheless help us make sense of what goes on in situations of actual confirmation in the sciences. I do not pretend, however, to supply answers or resolutions to certain classical paradoxes that have arisen in studying the logic of confirmation such as the 'Raven' paradox, the 'multiplication theorem' paradox or Goodman's 'Grue' paradox. Instead, my approach will be highly qualitative and informal in order to shed some light on the problem of how theories can be objectively confirmed.

2. POPPER'S APPROACH TO CONFIRMATION

Like Kuhn, Popper rejects the notion that we assess an hypothesis in terms of the degree of confirmation vested upon it by experimental results.[1] Again, this is because while it takes only one negative experimental result to refute an hypothesis, it would take an infinite number of positive instances to confirm it. Popper wants us to conclude that inductive reasoning plays no role in the establishment of our theories; contrary to traditional thought, there is no place whatsoever for induction in scientific methodology. Instead, he proposes the fallibilist viewpoint or the method of theory corroboration. Rather than basing our theory selection on the number of positive experimental results a theory generates, we try to develop hypotheses that are highly falsifiable (or 'bold') relative to possible rivals; and to the extent that the hypothesis in question resists falsification, we say it has achieved a degree of 'corroboration'. In other words, the more a theory risks, the more it gains each time it foils an attempt to falsify it. Suppose a neural-psychologist comes up with a theory of the brain which predicts that chickens will be able to solve mathematics problems if they are injected with a certain drug. Such an incredulous and surprising prediction would highly corroborate this theory if, indeed, it turned out to be true, and even more so if it worked for a variety of lower forms of animals such as fish and worms.

Popper's corroboration approach, of course, is based on the belief that there is a fundamental asymmetry between the verification and falsification of theories because a single experience or observation may serve to bring about the downfall of a particular scientific theory, but no finite number of observations can lead to its verification or confirmation. Hence, the logic of confirmation and refutation is not the same. (Recall how Kuhn denies that even refutation can take place in a single instance.) I will attack this claim to asymmetry below. But the more pressing issue concerns the overall outcome that our theories are *never* really confirmed by positive experimental results. On appearances alone, in spite of the fact that our theories are underdetermined by experience, we often speak of them as being highly confirmed. How can this be, according to Popper's corroboration view, especially in light of the fact that there are cases where only four or five instances of positive test results led the scientific community to accept a theory as being highly confirmed? This is precisely what happened in the case of the general theory of relativity. I believe that models of theory acceptance which lead to the result that our theories are not capable of confirmation are reminiscent of aeronautical engineers who devised a model intended to capture the aerodynamics of the flight of bees and, on the basis of their model, concluded that bees were too heavy to fly!

3. PROBABILISTIC PRINCIPLES UNDERLYING CONFIRMATION

In disagreeing with Popper, I do not mean to hold that theories can be absolutely experimentally confirmed. It is just that I believe it is possible to achieve a high degree of theory confirmation with only a few positive results, provided that they are the *right kind* of results. Popper's approach can not make sense of this feature of confirmation. What we need is a model that does.

One attempt to model the process of confirmation, and one which I feel is a step in the right direction, can be found in Salmon's *The Foundations of Scientific Inference*. Like many other philosophers who have done work in this area, Salmon proposes that we drop the notion of absolute theory confirmation, recognising that 'scientific hypotheses are proposed, not in an epistemic vacuum, but against the background of many previously accepted hypotheses and many that have already been rejected'.[2] Confirmation, here, is relative in that there is no test of an hypothesis without reference to rival hypotheses. Essentially, what a scientist does in testing out a theory is to 'bet' the said theory against its rivals. The question is what counts as winning the bet, i.e., what statistical principles serve to connect positive test results with the confirmation of a theory over its rivals?

Salmon's answer to the above question is based on Bayes's theorem. This theorem says that the probability of A given B equals the probability of A times the probability of B given A, divided by the probability of B:

$$\Pr(A/B) = \frac{\Pr(A) \times \Pr(B/A)}{\Pr(B)}$$

As a theorem that results from the multiplication and conditional probability rules, Bayes's theorem is fairly straightforward and innocuous; however, its application as a schema for theory confirmation is quite controversial. Substituting 'the probability that a hypothesis, H, is true given evidence, E' for $\Pr(A/B)$, we have

$$\Pr(H/E) = \frac{\Pr(H) \times \Pr(E/H)}{\Pr(E)}$$

which reads: the probability or confirmation that evidence E confers on hypothesis H equals the product of the initial probability that H is true and the probability that E is true given H, divided by the probability of E (independently of any and all hypotheses).

Now suppose we ran an experiment to test an hypothesis, H. We get a positive result, E. To what extent is H confirmed on the basis of E? The answer according to Bayes's theorem is that it depends on the initial probabilities of H, E/H and E. For example, if H is a highly dubious hypothesis, E does not confer very much probability in comparison to more likely hypotheses that equally predict E. Likewise, if E were an extremely likely event no matter what our hypotheses may be, then E's support for H diminishes. Of course, to the extent that H predicts E, E's occurrence confirms H.

One reason why Salmon thinks that Bayes's theorem captures the logic of confirmation is that the naive, deductivist view of theory verification and Popper's corroboration model – which have appeared as contraries – are beautifully unified as limiting cases of the above equation.[3] The deductivist view maintains that a theory is confirmed to the extent that it (exactly) predicts E. This appears as a special case of Bayes's theorem when $\Pr(E/H) = 1$. Popper's falsificationist theory occurs as a limiting case when $\Pr(E/H)$ is high while $\Pr(E)$ is very unlikely. This explains why the more a theory risks the more it gains in corroboration. Both the naive verificationist and Popper, then, have hit on true aspects of confirmation but, in themselves, they are incomplete schemata.

In spite of the attractiveness of Bayes's theorem as a model for the logic of confirmation, its application is surrounded with what may appear to be insurmountable difficulties. A cursory inspection of the theorem reveals that $\Pr(H/E)$ can not be calculated without knowing the *prior values* for $\Pr(H)$, $\Pr(E/H)$ and $\Pr(E)$. Fixing the prior probabilities for H and E turns out to be extremely difficult if not impossible, especially in actual working scientific situations. Take the general theory of relativity. Before any experiments were performed to test it, what was the likelihood that it was true? How does one even begin to answer such a question? (Suppose we get an answer which says that the theory is highly improbable in comparison to well-established classical theories of physics. Does this mean that a positive test result will not confer much probability on H because $\Pr(H)$ is so small? What if the other, rival theories cannot predict E? Even though H may be the best possible theory on the basis of comparison, it may not be highly confirmed according to Bayes's theorem. I will return to cases like these when I discuss the Principle of Maximum Likelihood.)

The same problem of fixing prior probabilities in an objective, non-arbitrary fashion holds for $\Pr(E)$ as well. It simply seems to be a practical impossibility to compute them on the basis of what we know. For example, we could, following Carnap, make an exhaustive list of all 'possible worlds' and set $\Pr(H)$ as the ratio of the number of possible worlds in which H is true over the total number of possible worlds. Even if the total number of possible worlds were finite, one wonders how such a ratio could be calculated without throwing in several unwarranted and arbitrary assumptions. Again, the same difficulty holds for setting $\Pr(E)$; until this problem is surmounted, one wonders how Bayes's theorem can be used to display the logic of confirmation without forcing us to conclude that we will never really know to what extent E confirms H.[4]

Salmon attempts to handle the above problem by using induction by enumeration as a basis for setting $\Pr(H)$ and $\Pr(E)$. In other words, our prior probabilities are determined by experience albeit such experience may be 'crude' in comparison to the subsequent Bayesian inference.[5] Although such a proposal may have an attractiveness for those of an empiricist bent, it seems to be oblivious to the fact that we have little or no previous experience concerning the probability of a *new* hypothesis. Isn't this what testing is all about in the first place? In other words, in order to use *any* experience to set prior probabilities for $\Pr(H)$ (and $\Pr(E)$), aren't we already in the process of confirming our hypotheses, and doesn't this mean that, in doing so, we are appealing to a principle of confirmation other than Bayes's theorem? The principle one might have in mind here, one currently used in statistics, is known as the Principle of Maximum Likelihood.

R. Fisher was the originator of the Principle of Maximum Likelihood. Put in simplest terms, it says: given a set of rival hypotheses, H_1, H_2, . . ., H_n and experimental evidence, E, the hypothesis that best supports the occurrence of E, i.e., the hypothesis according to which E is most likely, is the one for which E serves as evidence (or confirms).[6] Let us see how this principle works. Suppose we knew that an urn contained ten black and white balls but we did not know how many were black or white. We draw eight balls, without replacing them, and the sequence of drawings is WWBBBWBW. On the basis of the principle of maximum likelihood, since the hypothesis that there are five black and five white balls in the urn yields the highest probability for the above sequence of drawings, it has a higher degree of confirmation bestowed upon it by the above sequence than, say, the thesis that the urn contains three black and seven white balls, or any other hypothetical combination for that matter. We conclude that the above sequence best confirms the hypothesis that five balls are black and five are white. But this is only a confirmation of H *relative to other hypotheses* on the basis of an E *relative to other evidence or data*. Hence, H is not completely confirmed by the above sequence since it is compatible with the hypothesis that six balls are white and four black (or vice versa). We must make one more drawing to settle between the five–five and six–four hypotheses. Unlike Bayes's theorem, then, which must somehow have a catalogue of all the available hypotheses in order to determine $\Pr(H)$, the principle of maximum likelihood allows for the possibility that our competing theories are not necessarily exhaustive, that a new theory may come along someday which will show how E is even more likely. Until that theory comes along we will make do with the hypotheses and evidence we have at our disposal and assign the highest degree of confirmation to that theory which statistically best supports the data.

In light of the fact that we are pretty much in the dark when it comes to fixing prior probabilities to our theories and the outcomes of experiments, it appears to be more natural to use the principle of maximum likelihood instead of Bayes's theorem as a schema for the relationship between data and the hypothesis confirmed by the data. Not only is this principle able to avoid the problem of fixing prior probabilities, like Bayes's theorem, it, too, can render the deductivist view of confirmation and Popper's falsificationist approach as limiting cases. For this reason, I will incorporate the principle of maximum likelihood in my account of how an hypothesis can be highly confirmed by just a few, key positive experimental results.

4. THE LOGIC OF A CRUCIAL EXPERIMENT

Crucial experiments are experiments that are designed in such a way that their results serve to decide between rival theories or hypotheses. Somehow, the data of a crucial experiment simultaneously indicate the truth of a given theory while entailing the denial of its contrary. Now there is much controversy over the question of whether or not crucial experiments can really exist at all in scientific practice. I have noted that one outcome of Kuhn's analysis of theories and his incommensurability thesis is that there are no such things as crucial experiments (between fundamental theories or paradigms), while the positivists and formalists claim that they exist all the time. Before this issue can be settled, I shall first try to establish the nature of a crucial experiment in terms of its logical structure. In doing so, I shall also call into question Popper's claim that there is a logical asymmetry between verification and falsification of theories.

It is well known by now that a theory or hypothesis does not lead to its own test. What we need in order to generate observable predictions are some auxiliary hypotheses which serve as links between theoretical properties (or entities) and observable features in nature. Recall the discussion about rival geometrical theories of space in Chapter 2. In order for us to test the hypothesis that space was Euclidean (or non-Euclidean), we had to combine the Euclidean hypothesis with an auxiliary hypothesis about the behaviour of rigid rods under transport, for example; the length of a rod is preserved under transport or its length is not preserved under transport because it is affected by 'universal forces'. Let us call these theories measuring theories, θ_m. So when our theory to be tested, θ_1, is combined with θ_m, their combination will imply an observable result, O:

(1) $(\theta_1 + \theta_m) \rightarrow O$.

Since it is assumed in testing θ_1 that θ_m is true, if O turns out to be false, then it logically follows that θ_1 is false, i.e., our theory is falsified. However, as any falsificationist has insisted, if O is true, θ_1 can still be false; so the truth of O does not necessarily confirm or verify θ_1. And, as Popper insists, no finite number of positive observations can lead to the truth of θ_1 in the same way that one single negative case can lead to its falsification. Hence, we have Popper's asymmetry between verification and falsification.

But the above view overlooks an often found possibility of combining *another*, rival theory, θ_2, with θ_m yielding NOT-O as a prediction:

(2) $(\theta_2 + \theta_m) \rightarrow$ NOT-O.[7]

θ_1 and θ_2 are rival theories in that while they can both be false, they can not both be true. And, of course, a crucial experiment between them on the basis of (1) and (2) assumes that (3) θ_m is true. (Notice that in order for θ_m to be combined with θ_1 to yield O, logic dictates that they have concepts in common. The same for θ_m and θ_2, which means that θ_1 and θ_2 would be rival theories that share concepts – something Kuhn would have vigorously denied, according to his early position.) Now if we are to avoid Kuhn's charge that crucial experiments are guilty of circularity because their interpretation presupposes the very paradigm they are intended to establish, then the truth of θ_m must be logically independent of θ_1 and θ_2.

So far, as things stand, an experimental result of O or NOT-O can not as yet settle between θ_1 and θ_2; we need one more premiss pertaining to the logical relationship among rival hypotheses. I said that rival hypotheses could not both be true but they *could* both be false. Yet, when scientists perform a crucial experiment intended to decide between two rival hypotheses, they are of course assuming that one of them is true. Otherwise, why perform the experiment? Is the assumption that either θ_1 or θ_2 is true merely an act of scientific faith, a breakdown in objectivity which parts from empirical tradition? I think not. Rather, such assumptions, in my opinion, reflect deep conceptual truths *which are common to θ_1 and θ_2*. While propositions like 'either θ_1 or θ_2 is true' are in principle falsifiable, falsifying them is not so easy a thing to do, in that doing so will have conceptual implications of a far-reaching nature. That they may be false is another way of saying our rival theories are not exhaustively so, something which we can not deny. But emphasising their falsifiability can lead us to overlook the conceptual power of these claims, as many philosophers have.

The fact of the matter is that propositions which state (or entail) that one of a set of rival hypotheses is true are not like ordinary disjunctive claims such as 'Either John went home or to the shop.' Discovering that such a proposition is false would not leave us with any conceptual scars but learning that the alternative presented by two rival theories is false would do so in the sense that it would call for a revision of our conceptual scheme in the same way that anomalies worked to place conceptual stress on a given scientific paradigm. The falsification of disjunctions such as 'Either the Earth moves or it doesn't', 'Either the speed of light depends on the speed of the inertial frame of the light source or it doesn't', 'Angular momentum is either a quantised or a continuous quantity' leaves us 'traumatised' because it is not an easy matter to 'bust' these dichotomies. If angular momentum is a quantity that is neither continuous nor discontinuous, then what kind of a quantity is it? Maybe it would leave us with the perplexity that angular momentum is not really quantifiable at all in the same way natural philosophers in the past such as Galileo, Descartes and Locke treated colours, sounds, smells, etc., as 'secondary qualities' which were not worthy to be classified as real or inherent qualities of things. If so, contemporary physics, classical or quantum, would be left in a quandary. Consider 'Either the speed of light depends on the speed of the inertial frame of its light source or it doesn't.' What would it be like to bust this dichotomy? Well, at one time, the scientific community believed that light 'moved instantaneously'. That would do it, but we would then be very hard-pressed to explain time delays between a variety of astronomical events and their subsequent observation on Earth. It would be even more difficult if not impossible to find a system of kinematics which maintained that light moved instantaneously or a conception of light which made sense of such a feature. Likewise, if the earth is neither moving nor at rest, then what? This one is 'easier' than the other two. All it takes is to replace absolute motion with relative motion along with some principle of relativity. Neither the Earth nor the sun is at rest. (But what about: either the Earth rotates or it does not?)

It can be seen how breaking a scientific dichotomy requires imagination and the systematic rearrangement of our conceptual repertoire. Although I have spoken of scientific dichotomy in terms of rival theories, we may also speak of an *aspect* of one theory being dichotomous to an *aspect* of another theory. As we shall see, very often, only part of a theory is bet against another in a crucial experiment but the logic is the same: the experiment presupposes that θ_1 and θ_2 are dichotomously related, i.e., (4) θ_1 or θ_2 is true (where (4) expresses or entails a scientific dichotomy). From (1)–(4), it follows that (5): θ_1 is true if and only if O is true; and θ_2 is true if and only if NOT-O is true.

We can call experiments or observations which yield statements like those in (5) 'Yes–No' experiments, because θ_1 (and θ_2) is true or false depending on whether O or NOT-O is true. Thus, in cases as these, the truth of O can decide between θ_1 and θ_2 without having to resort to artificial, internal criteria such as simplicity, clarity, problem-solving ability, and so on. I think that (1)–(5) captures the logic of a crucial experiment. If it does, however, Popper's claim that there exists an asymmetry between verification and falsification breaks down, for according to the above schema, a single refuting instance of one hypothesis is equivalent to confirming the other and vice versa. So, if we simply reverse the negation sign for O in (1)–(5), confirmation and refutation are then reversed. What more could we mean by the symmetry of confirmation and refutation?

Now let us turn to some actual cases of a crucial experiment. This first case was used to test between classical and quantum mechanical conceptions of angular momentum, known as the Stern–Gerlach experiment. A beam of silver atoms is sent through a very inhomogeneous magnetic field. After the beam passes through the field, it is deposited on a glass plate. According to electrodynamics, if the atoms of the beam have an angular momentum (called spin), then a magnetic force will be exerted on each atom as it passes through the field, a force that will depend on the amount and direction of angular momentum. The effect of this force is to deflect the path of the atom. Now, and most important, *this is true independently of either quantum mechanics or classical theories of motion*, and this served as our θ_m. Commonsensically, it would appear that each silver atom could spin at just about any rate, i.e., silver atoms could take on a continuous range of angular momenta. This is what classical mechanics, θ_1, allows. Anything so commonsensical automatically becomes suspect in quantum theory, so it should not be surprising that quantum theory, θ_2, predicts that the angular momenta of the atoms are 'quantised', i.e., they do not take on a continuous range of values; rather there are large 'gaps' between amounts of spin. So when θ_1 is combined with θ_m, what should we expect to observe on the glass plate after sending a beam of silver atoms through the magnetic field? The atoms should deposit on the glass in the form of a continuous smear, as in Figure 6.1.

Glass Plate

FIGURE 6.1: *Deposit of silver atoms after passing through a Stern–Gerlach apparatus*

On the other hand, when θ_2 is combined with θ_m, since angular momentum is quantised, the smear is discontinuous and should look like Figure 6.2.

Glass Plate

FIGURE 6.2. *Pattern of a beam of silver atoms of quantised angular momentum*

When the experiment was actually performed, a split beam was observed. By the logic of (1)–(5), that aspect of quantum theory which deals with angular momentum was confirmed while the rival, classical mechanical view on angular momentum was simultaneously refuted. (As I have already mentioned, this assumes that angular momentum is either continuous or discontinuous. Again, what would it be like to deny such a scientific dichotomy?) It is interesting to note that although this result was expected by the scientists who devised and performed this crucial experiment, actually observing the discontinuity occur was almost magical and it was nevertheless quite a shock to them.

Consider another example. In spite of Galileo's lengthy polemics against the geocentric astronomy and his defence of the heliocentric view, he was hard-pressed to supply the reader with a genuine or feasible crucial experiment to decide between these fundamental rival theories. Well, given the astronomical facts, either (4) the Earth is at rest *vis-à-vis* the sun, θ_1, or the sun is at rest and a spinning Earth circles it, θ_2. It seems that we have no other choice. The reason why Galileo had such a difficult time providing the reader with a crucial experiment was that he had lacked a θ_m which could be suitably combined with θ_1 and θ_2 to lead to opposite observable results, something Galileo, of course, could not really help, as Newton's laws came some time later. However, combining the view that the Earth is at rest with Newton's laws implies that the path of a rocket sent from the north pole to the equator will be perfectly straight. But the path of a rocket that travels above a rotating frame of reference, as it is with a spinning Earth, will ultimately be observably deflected longitudinally. (This phenomenon is referred to in physics as a Coriolis acceleration, which is the result of motion measured on a rotating frame of reference.) We thus have a crucial experiment to decide between the two rival astronomies. (Notice that Newtonian mechanics does not entail the truth of either θ_1 or θ_2.) Actually, there is a much simpler experiment that can be performed but one based on the same principles: fill up a bath tub with water and then pull out the plug. θ_1 and θ_m predicts that the water will spiral down the drain in any direction while θ_2 and θ_m says that the water will spiral clockwise or counterclockwise down the drain, depending on whether the bath tub is located in the Northern or Southern hemisphere of the Earth. This crucial experiment is far less costly than the other but the reasoning is basically the same.

One of the most famous examples of a crucial experiment was the Michelson–Morley measurement of the speed of light along different directions to the motion of the Earth. The classical view of light said that light waves moved in a medium (called the ether), and that like a sailboat moving on a river, the speed of light will vary, depending on whether it is moving 'downstream', i.e., in the direction of the Earth's motion, or moving perpendicularly to the Earth's motion. Thus, classical theory, θ_1, predicts that the journey of a beam of light back and forth along the direction of the Earth's motion will take a different time than one perpendicular to the Earth's motion in the ether. Now one of the more surprising claims of the special theory of relativity, θ_2, is that the speed of light is independent of the light source's frame of reference. Special theory of relativity thus predicts that the journey of the beam will take exactly the same amount of time no matter what the direction of the beam is relative to the Earth's motion. The trick is how such a difference in time can be measured in view of the fact that the distance covered by the light is so small. Michelson and Morley solved this problem by utilising a principle found in optics, one which, once again, is true independently of the rival theories. The principle of superposition says that two different waves can 'add up' to form a single wave which is a linear combination of the two original ones. If the waves are in phase, they will form a vivid, coherent pattern while out of phase waves will cancel out, forming a 'fuzzy', faded pattern. This, then, is our θ_m. An interferometer can be used to detect coherent and incoherent wave patterns. Now take a beam of light, split it into two beams, send one beam on a journey parallel to the Earth's motion while the other journey is perpendicular. Have the two beams converge back again through an interferometer at the end of their journeys. Then rotate the entire apparatus ninety degrees and do the experiment again. If it took light different amounts of time to traverse an equal distance under one orientation with respect to the Earth's motion in the ether as opposed to one rotated ninety degrees, there should be a phase shift in the interference pattern of the returning light beams once the apparatus is rotated. So, θ_1 and θ_m predicts that there will be a phase shift of the light pattern after rotation while θ_2 and θ_m predicts that no such shift will take place. Much to the surprise of the scientific world, no phase shift was observed under a variety of circumstances. The special theory of relativity's fundamental claim, that the speed of light was the same no matter what frame of reference in which it was measured was confirmed while another commonsensical classical conception about light was refuted.

The above experimental result was so shocking that a serious attempt was made to develop a rival theory which explained it away. It was proposed by Lorentz that measuring instruments which moved along the direction of the Earth experience a kind of compression, causing them to shorten by a miniscule amount (as a function of velocity) but an amount which equally compensates for the lack of discernible difference in the light patterns. In other words, even though the speed of light is faster in one direction, because the measuring instruments along that direction contract, the time it takes each beam to make the trip to and from the mirror ends up being the same! Hence, we should expect no change in the phase pattern when they reach the interferometer under different rotations. The Lorentz–Fitzgerald contraction can thus be used to explain away the results of the Michelson–Morley experiment, that is, in such a way to void a confirmation of the special theory. (Essentially, it denies that $(\theta_1 + \theta_m) \rightarrow O$.) The fact that *ad hoc* moves such as the Lorentz–Fitzgerald contraction in the case of the Michelson–Morley experiment are always available to rescue a theory from refutation via a crucial experimental result has inspired Kuhn to deny that any kind of experimental results can really decide between rival paradigms. Hence, his thesis that fundamental theories are incommensurable. It does look as though *ad hoc* moves are always available to challenge the results of a crucial experiment but such moves are made at a terrible empirical price, which is something Kuhn overlooked, as we will soon see in the next section.

5. TESTING THEORIES ACROSS CATEGORIES

It is my contention that theory confirmation, in practice, utilises the above logical model for a crucial experiment in combination with the principle of maximum likelihood. By this means, I will show how a high degree of confirmation of a theory can be obtained with just a few, crucial experimental results, something Kuhn and the falsificationists said could not be done. Let us return to Salmon's point that confirmation does not take place 'in an epistemic vacuum' but instead by 'betting' a hypothesis, so to speak, against rivals. Using the principle of maximum likelihood, the degree of confirmation allotted to a hypothesis is a function of how high a probability it assigns to E in comparison to other, rival hypotheses. Suppose we want to know the extent of confirmation that experimental results E confer upon a group of four hypotheses, $\theta_1-\theta_4$. According to the likelihood principle, we calculate the probability of E given H, $\Pr(E/H)$ and represent it by this table:

$\Pr(E/H)$	θ_1	θ_2	θ_3	θ_4
	.05	.65	.25	.05

By the above, θ_2 is the best confirmed by E because $\Pr(E/\theta_2)$ is higher than the others.

But things are not quite that simple when it comes to pitting one's theory against its rivals. One very important feature about scientific theory, which has often been overlooked, is that the confirmatory and explanatory powers of a theory depend, in part, on how many-faceted it is, i.e., on the number of different aspects of nature the theory happens to cover.[8] Some theories may be extremely narrow in scope by covering only a single aspect of nature while others may have something to say about a broad range of things. For example, the hypothesis that space is Euclidean does not, in itself, say anything about time, matter and gravity as the general theory of relativity most certainly does, including things about the nature of space. So there are three extremes when it comes to comparing rival theories: θ_1 may be a single-aspect theory which is tested against θ_2 with respect to that aspect; several aspects of θ_1 are systematically tested against the corresponding parts of θ_2; and several aspects of θ_1 are tested against separate, independent theories, $\theta_2, \theta_3, \ldots, \theta_{n+1}$, where each theory covers a corresponding aspect of θ_1. (It is the third extreme in which I am most interested, because it best fits the realist–positivist dispute which will be covered in the next chapter.) For example, suppose we had an urn filled with objects that were either black or white in colour and square or circular in shape. An instance of the first type of comparison would be θ_1 saying that half of the objects were black and one half were white while θ_2 could claim that one-third are black and two-thirds are white. θ_1 and θ_2 are only about colour. On the other hand, we might hypothesise, θ_1, that half of the objects in the urn are black squares and half are white circles; θ_2 might be one-third black circles and two-thirds white squares. θ_1 and θ_2 cover both aspects of shape and colour. Finally, θ_1 might be a hypothesis saying that half of the objects are black squares and half are white circles. Now suppose we had two theories, θ_2 and θ_3, which *independently maintained* things about the colours and shapes of the objects; for example, θ_2 says that half of them are white and the other half are black while θ_3 claims that half of them are squares, half are circles. Thus θ_1, which says something about both the colour and shape of the objects in the urn, would go against the separate theories of θ_2 and θ_3.[9] Once we define how our theories are to be compared in terms of their aspects, we can begin to select balls from the urn in order to determine which of the three theories is best confirmed.

Let us examine the third way of comparing rival theories in more detail. Suppose we wanted to test a many-faceted theory with several, independently held single-faceted theories. In this case, we should have a crucial experiment corresponding to each aspect of the many-faceted theory which is bet against its rivals. So if θ_1 has n-number of aspects we ideally should devise n crucial experiments with n results:

$$(\theta_1 + \theta_{m1}) \rightarrow O_1 \qquad (\theta_1 + \theta_{m2}) \rightarrow O_2 \ldots, (\theta_1 + \theta_{mn}) \rightarrow O_n$$

$$(\theta_2 + \theta_{m1}) \rightarrow \text{NOT-}O_1 \quad (\theta_3 + \theta_{m2}) \rightarrow \text{NOT-}O_2 \ldots,$$
$$(\theta_{n+1} + \theta_{mn}) \rightarrow \text{NOT-}O_n$$

$$\theta_1 \text{ or } \theta_2 \qquad\qquad \theta_1 \text{ or } \theta_3 \qquad\qquad \ldots, \theta_1 \text{ or } \theta_{n+1}$$

Each experiment has the logic of a crucial experiment outlined above. Because θ_1 is to be tested against a series of rival hypotheses, beginning with θ_2, θ_{n+1} denotes the number of the hypothesis which is to be compared with a particular aspect of θ_1. O_n naturally stands for the data of that crucial experiment which serves to test between θ_1 and θ_{n+1}. The first subscript in θ_{mn} lets us know that θ, here, is a measuring theory relative to a crucial experiment while the second subscript simply denotes the number of that experiment. The arrow reads 'given that the conjunction of θ_n and θ_{mn} is true, the odds that data O_n are true are such and such'. This means that O (or NOT-O) need not be a deductive prediction from $\theta_1 + \theta_{mn}$; on the contrary, $\theta_1 + \theta_{mn}$ simply assigns a probability for O_n's occurrence or its probability distribution when several O_n-type measurements are made. This is in keeping with actual scientific practice, for while our theories may assign 1 or 0 to O_n this is just an ideal which is never fully realised in experimentation in light of the fact that an experimental apparatus is never perfect, that the information it gives is never completely 'noise' free. In reality, then, when a measuring theory is combined with θ_n, we should not expect to get a perfectly sharp prediction of O_n. This is built into information-processing theory and the theory of instruments.

With the above qualification in mind, let us see how the combination of n-crucial experiments and the principle of maximum likelihood works for the third case of theory comparison, namely, when various aspects of θ_1 are tested against independently maintained theories, $\theta_2, \theta_3, \ldots, \theta_{n+1}$. Suppose each crucial experiment between θ_1 and θ_{n+1} leaves very little predictive difference between them:

$$(\theta_1 + \theta_{m\,1}) \to .6O_1 \qquad\qquad (\theta_1 + \theta_{m2}) \to .6O_2 \ldots$$
$$(\theta_1 + \theta_{mn}) \to .6O_n$$
$$(\theta_2 + \theta_{m\,1}) \to .4O_1 \text{ (or .6NOT} - O_1) \quad (\theta_3 + \theta_{m2}) \to .4O_2 \ldots$$
$$(\theta_{n+\,1} + \theta_{mn}) \to .4O_n.$$

There is not much difference, here, between θ_1 and each θ_{n+1} since θ_{n+1} predicts NOT-O_n with just .6 probability or O_n with .4 probability. According to the principle of maximum likelihood, each crucial experiment which yields O_n lends a degree of confirmation to θ_1 over its corresponding θ_{n+1} rival, but not very much confirmation since the difference in each experiment is only .2. That is only if we consider the results of each crucial experiment *in isolation* to the others, something that should not be done. Now what happens if we apply the principle of maximum likelihood to *all* the crucial experiments, which are n in number, together? First of all, we want to determine what statistical support our theories confer on the *conjunction* of experimental results, E_n. Once we know that, we can then compare to see which theory or theories lend the highest degree of probability to the total E_n.

The problem, of course, is to determine how our theories lend statistical support to conjunctions of results E_n to begin with. Since, in the above case of comparison, $\theta_2, \theta_3, \ldots, \theta_{n+1}$ are *independently maintained* theories, each one covering a specific aspect of nature, we can not assign a probability for a combination of E_n results on the basis of a single θ_{n+1} alone since θ_{n+1} has nothing to say about other aspects of nature. What we must do in order to rectify this situation is to form another theory which is the *conjunction* of $\theta_2, \theta_3, \ldots, \theta_{n+1}$, i.e., the statistical support our rival theories confer on the total E_n is a function of the support the individual theories lend to individual E_n. Since the experimental results, E_n, are independent events according to θ_2, θ_3, . . . , θ_{n+1} (because they are independent theories), the probability of the combination of E_n given the conjunction of θ_2, θ_3, . . . , θ_{n+1} is simply the product of the probability each individual theory assigns to its corresponding E_n. Not so, θ_1! In this case, assigning a probability for the total E_n is not done by taking the product of each individual $Pr(E_n)$ because, *according to θ_1, the E_n experimental results are not independent events*. This is illustrated in the coloured shape example when we compare the hypothesis that, θ_1, half of the objects in the urn are black squares and half are white circles with the rival hypotheses that, θ_2, half are white and half are black and, θ_3, half of them are square while the other half are circles. If we want to calculate the probability of getting both a square and black joint experimental result, θ_2 and θ_3 do not do this alone but they must be conjoined to yield .25 for a black–square result. However, θ_1 assigns a probability of .5 for the same joint result because according to θ_1, shape and colour are not independent of each other. (In the next chapter, I show how the probabilities theories assign to joint experimental results actually depend on the ontology expressed in the theory, and I will go into the exact nature of this dependence.)

Now we can come back to the above series of crucial experiments where each individual result lent only a small degree of confirmation to θ_1 (the difference being .2 in each case). What happens when we take all the results together? The statistical support θ_1's rivals lend to the conjunction of E_n results goes down *geometrically* in comparison to θ_1:

$$Pr(\text{Total } E_n/H) \frac{\theta_1 \quad \theta_2 \text{ and} \theta_3 \text{ and } \theta_4 \ldots \text{ and } \theta_{n+1}}{.6 \quad .4 \times .4 \times .4 \ldots \times .4}$$

So if we ran only five crucial experiments across categories which cover five different aspects of nature, even though θ_1 does not fare that much better than each individual θ_{n+1}, it beats them all by .6 to .01 when the five experimental results are taken together! The upshot of all this is that if we want to run experiments that lead to a high degree of confirmation (or refutation), instead of running many experiments of a single type, it is much more efficient to test our hypotheses across categories. And this is why the general theory of relativity is granted an extremely high amount of confirmation on the basis of a relatively small number of experiments; for this theory has something to say about many facets of nature while it is tested via a crucial experiment against rival theories which are not as many-faceted. Euclidean geometry does not have anything to say about time; the classical picture of time is independent of gravity; classical dynamics stands independently of geometry – and so on. But, according to the general theory, all of these things are aspects of the same, deeper phenomenon, namely, a geometricised space–time continuum of events.

Another important thing that the above cases of confirmation show is that it is quite possible to bring about a high degree of theory confirmation without necessarily having to falsify its rivals, as Popper maintains, for a difference of only .2 between θ_1 and each of its rivals, in itself, should not falsify θ_1's rivals (which individually assigns a .4 probability for each E_n). In fact, θ_1 need not even lead to a higher probability for E_n in relation to each of its rivals! Again, if θ_1 says that half of the objects in the urn are black squares while the others are white circles, θ_2 says that half of them are black and half are white while θ_3 claims that half are square and the other half are circles, a very interesting situation arises, one that is often found in the sciences. Very often, a theory may be more confirmed by the total evidence simply because it covers more aspects of nature than its rivals.[10] Notice in the above case, if we restrict the evidence to each, separate category, θ_1, θ_2 and θ_3 are equally confirmed. Not only that, θ_2 and θ_3 are true in their own right! Nevertheless, if we were to measure the frequency of black *and* square objects and find that half of our observations of both shape and colour yield black and square, then by the principle of maximum likelihood, θ_1 is still more confirmed than the combination of θ_2 and θ_3; the latter gives statistical support of .25 for the joint occurrence of black and square events while the former predicts .5. This situation most closely resembles the approximation view of scientific progress which was discussed in the last chapter. So a very important reason for testing our theories across categories is that even if the evidence tends to confirm a theory when its scope is restricted to a single aspect of nature, it may turn out that such a theory may go wrong when tested across several aspects. Here is a clear-cut case of confirmation where falsification does *not* take place.

I promised that I would show how both the D–N model of theory confirmation and the Popperian falsification view can be incorporated into my model of confirmation as two limiting cases of testing across categories. There is not much to do in the case of the D–N approach except to show how restrictive that view can be. Confirmation occurs when E_n is given a likelihood of one according to θ_1 and zero according to its rivals. Strictly speaking, confirmation need not even involve comparison of θ_1 to its rivals or even testing across categories for that matter if we accept the D–N view. Suppose we devised a crucial experiment to decide between θ_1 and θ_2 with respect to a particular aspect of nature where $(\theta_1 + \theta_m)$ predicts O_1 with a probability of one while $(\theta_2 + \theta_m)$ predicts NOT-O_1 with a probability of one. Suppose an experiment yields O_1, confirming θ_1 and refuting θ_2. Should we now go on to test θ_1 with respect to other aspects of nature? Ironically, there would not be any need here to perform more crucial tests of θ_1 with respect to these other aspects because in the limiting deductive case where $(\theta_1 + \theta_m)$ predicts O with probability one, all other θ_1's rivals are truth functionally (or materially) equivalent to θ_2. Showing θ_2 to be false with a probability of one is tantamount to showing that all other, equivalent theories are false. This follows directly from the logic of a crucial experiment and the transitivity of the equivalence relation. We know that the truth of θ_1 and NOT-θ_2 is equivalent to O_1; this holds for any crucial experiment, n. So it follows that the denial of any rival theory is equivalent to O_1. Thus, a single crucial experimental result is enough to take care of all of θ_1's rivals. Now this inference does not go through if $\theta_1 + \theta_m$ predicts O_1 with a probability less than one. This is because, according to the principle of maximum likelihood, transitivity across categories of probabilistic prediction and confirmation does not go through, i.e., if E confirms θ_1 and θ_1 is materially equivalent to θ_2, it does *not* at all follow that E also confirms θ_2. If the hypothesis that half of the objects in the urn are black squares while the others are white circles is true, then the sub-thesis that half are black and half are white is truth functionally equivalent to the sub-thesis that half are squares and half are circles. While this is so, experimentally finding out that the frequency of black and white objects drawn out of the urn is about equal does *not* serve to confirm the second sub-thesis because that hypothesis does not lend any statistical support for anything about colours. It may very well be that a subsequent experiment yields a ratio of square to circle which does not equal 1/2, refuting θ_1. Hence, in order to get the complete story of θ_1's confirmability, we must examine *all* of its *n*-facets by gathering E_n bits of experimental data. (I strongly suspect that the best way to handle the raven paradox – namely, since 'All ravens are black' is logically equivalent to 'All non-black objects are non-ravens', finding a white shoe is evidence for the truth of *both* propositions – is to deny the transitivity of statistical support.) Needless to say, this deductivist situation never arises in scientific practice because the ideal of predicting O (or NOT-O) with a certainty of one doesn't arise. So we have to test θ_1 against theories that cover rival aspects of nature anyway.

How is Popper's falsificationist approach rendered a limiting case of my model of confirmation? Recall how Bayes's theorem accomplished this. The formula for $\Pr(H/E)$ collapsed into Popper's model when $\Pr(E)$ was extremely low and $\Pr(E/H)$ was very high or approached one. But how do we know the $\Pr(E)$ is extremely low? This was our problem of fixing prior probabilities. The most natural reason we can give for believing $\Pr(E)$ to be very small, on the basis of information at hand, is that other, previously well-confirmed hypotheses say that E is highly unlikely, which has nothing to do with Bayes's theorem. This is another way of saying that while E may be highly supported by θ_1, it is given very low support by θ_1's rivals. This is just one extreme way of comparing theories, the Popperian or falsification way, but there are other ways to compare.

I have been arguing all along that the best way to test a theory is to do so across categories. There is another, excellent reason why we should test across categories, even if the θ_1 in question makes an almost perfect prediction relative to a single aspect of nature. As I have already pointed out, there is always the possibility that the advocates of a rival theory, θ_2, will explain away the experimental results that favour θ_1 by adding a suitable *ad hoc* hypothesis to θ_2 which equally predicts O. I have stressed that the possibility of such a move has served as a basis for the Kuhnian position that experimentation can not ultimately decide between fundamental rival theories. Kuhn feels that if we can always preserve our theory by the device of adding an *ad hoc* hypothesis, our refusal to do so is *not* an empirical matter but because the scientific community teaches us (by imprinting certain values in us) not to do things like this. As a way of illustrating this type of reasoning consider the traditional arguments that philosophers of science such as Schlesinger have given against the utilisation of an *ad hoc* hypothesis in order to preserve one's theory in the face of contrary data.[11] Suppose we wanted to insist that the Earth was flat. As Copi points out, we can always maintain such an absurd hypothesis no matter what.[12] (Again, our theories are underdetermined by experience.) For example, the reason why travelling due east always takes one back to one's starting point is because there is a force field which 'bends' motion on a flat Earth into a circle; the reason why ships appear to be lowering on the horizon as they go out to sea is that light waves bend; pictures from satellites and of eclipses are really a hoax; and so on. We can always be inventive enough to make up these hypotheses but, as Schlesinger points out, doing so amounts to adopting an ever-increasing number of *ad hoc*, unestablished hypotheses. If we had our choice between accepting the hypothesis that the Earth is round or the hopelessly complex set of unestablished hypotheses about mysterious forces, bending of light rays, etc., including the Flat Earth hypothesis, we are, of course, rational enough to choose the former over the latter for reasons of simplicity alone. While I completely agree with this argument against preserving a theory by making *ad hoc* moves, the Kuhnians will most certainly embrace Schlesinger's reasoning as another way to emphasise that we do not use empirical reasons to accept or reject fundamental theories (or any theory for that matter). Simplicity, they will say, is *not* an empirical consideration but a metaphysical one, one which is highly valued by the scientific community. So the argument for incommensurability goes.

My reply to the above argument is that putting simplicity and other socio-political considerations aside, the Kuhnians are still wrong because the 'simpler' theory is also more highly confirmed on the basis of the evidence than its rivals plus corresponding *ad hoc* theories. This is so in spite of the fact that the latter predict the same experimental results as the former, simply because, in the latter case, the *ad hoc* theories are *independently* maintained. In order to bring this out, apply my model of confirmation to the case in which θ_1 is bet not against its rivals but each rival combined with a suitable *ad hoc* hypothesis that predicts O_n with a probability equal to that of θ_1. Suppose $\theta_1 + \theta_{mn}$ predicts O_n with a probability of .9. So each revised rival theory θ'_{n+1} also predicts O_n with a probability of .9. Even so, when it comes to statistically supporting the total evidence, i.e., the conjunction of E_n, the statistical support for the total evidence supplied by theories again goes down geometrically in comparison to θ_1:

$$\text{Pr}(\text{Total } E_n/H)\,\frac{\theta_1}{.9}\quad\frac{\theta'_2 \text{ and } \theta'_3 \text{ and } \theta'_4 \ldots \theta'_{n+1}}{.9 \times .9 \times .9 \ldots \times .9}$$

In other words, as we add *ad hoc* hypotheses across categories in order to preserve our theories against θ_1, the confirmation of θ_1 ever increases at a geometrical rate. For example, if we add five, independent *ad hoc* hypotheses, θ_1 nevertheless better supports the total of experimental results than its rivals by .9 to .59! The conclusion is that the data better confirm θ_1 than its modified rivals, and that genuine empirical considerations are ultimately at work for accepting a theory over its rivals. This should at last put an end to the view that the 'taste' of the scientific community has the final say when it comes to theory acceptance.

Oh yes, the Kuhnians may insist that even the way we draw the line when it comes to enough difference in statistical support to warrant the acceptance of θ_1 over its rivals is a reflection of the 'metaphysical' values of the scientific community. We have come back, this time, to Kuhn's 'values of accuracy' thesis which claims that what we count as an accurate enough result to confirm our hypothesis depends on the values of the scientific community. I have already maintained that this claim equivocates between standards of accuracy being values in themselves as opposed to being determined by other values,[13] especially social-political ones. The Kuhnians tend to favour the latter interpretation, maintaining, for example, that the more negatively valued an hypothesis the more rigorous our standards of confirmation. For example, some have argued that if the truth of an hypothesis suggests policies that lead to disastrous social-political consequences, we take a much closer, more rigorous look at the experimental data before we assign any confirmation. If an experimental failure leads to great harm we are especially careful about the accuracy of our predictions of the experiment's outcome.

While I agree with the above to some extent, I do not believe it shows that our standards of accuracy concerning prediction and confirmation are a function of values in the sense that the Kuhnians intend. I think that accuracy in cases such as these is still a function of the nature of the objects, our theories, including our theories of instrumentation. What I am getting at is this. The Kuhnians have simply confused standards of accuracy and confirmation of our theories with those values that enter into the *implementation* of our theories. With regard to the latter, our theories are dumb in the same way that they do not contain their own tests but must eventually be combined with measuring theories in order to generate observable test results. In other words, while it is one thing to say that a theory has been accurately confirmed, it is indeed quite another to conclude that it should be put into practice. No doubt, the latter judgement demands that we carefully weigh the benefits of implementing such a theory, and even if we think that said theory is true, we might very well hesitate to use it for fear of the consequences. But this means that factors other than evidence and calculation enter into our practical decisions, which should not be very surprising. With the risk of oversimplifying things, there is a way mathematically to express the above distinction between standards of accuracy (and confirmation) and considerations concerning whether a theory should be implemented. It is known as the *expectation value* of an hypothesis: the expectation value is the product of the odds of an hypothesis being true and the benefits (or pay-off) of it coming out to be true. While the likelihood of the truth of the hypothesis has nothing to do with values, the assignment of benefit most certainly does. So we can easily explain the above two cases without having to resort to the dubious claim that our standards of accuracy are clouded by our social-political values. Suppose someone invites you to play a game of chance, using a roulette wheel which has one hundred slots. The game is this. Your opponent bets that the ball will land on a particular number. If it lands on any other number, you will win the bet and be paid five dollars. On the other hand, if the ball just happens to land on your opponent's number, you lose and you pay by forfeiting your life. Of course, you will not consent to play such a game even though you *objectively* know that the odds are 100–1 in your favour. Admittedly, your decision was based, for the most part, on your system of values and, perhaps, a fear of death, but this does *not* in the least affect the odds of the game. The same holds true for standards of accuracy and confirmation as opposed to how we should go about testing and implementing our theories. The former type of question is what this chapter has been all about; the latter concerns not just the scientific community but everyone, as a practical, everyday problem of what ought to be done. If a theory is to be refuted at all, no matter how obnoxious the consequences of putting it into practice may seem, it should be refuted by appealing to good, scientific methodology alone, and not because its truth *might* lead to immoral actions. After all, no theory is safe from that possibility!

In conclusion, I have argued that despite the fact that our theories are underdetermined by experience, they are still open to a process of confirmation which is objective in nature. Scientific progress can be grounded by experimental reasoning without begging questions of values. Once the confirmation of a theory is relativised to a background of rival theories, and once the logic of theory comparison between aspects of rival theories by means of crucial experiments is fully understood, we are then able to apply the principle of maximum likelihood in order to scale the degrees of confirmation among theories relative to the experimental data at hand. Although this process of confirmation is intended to reflect a fundamental value of the scientific community, namely that of objectivity when it comes to accepting or rejecting a hypothesis relative to other hypotheses, I see no reason why the above process of confirmation should be subject to the whim of the social-political values of a particular scientific community or paradigm.

Finally, the above model of confirmation holds the key to a very important clue on the logic and nature of scientific explanation. We have seen that part of a theory's power to gain a relatively high degree of confirmation *vis-à-vis* its rivals lies in its ability to unify a vast range of data or phenomena across categories, i.e., to reduce the independence of a number of phenomena. I shall now go on to argue that herein lies a theory's power to explain.

NOTES AND REFERENCES

1. But for different reasons. Representative passages on Popper's work in this area can be found in *The Logic of Scientific Discovery* (London: Hutchinson, 1968) especially in sections 31–46, 80–5 and in his *Conjectures and Refutations* (London: Routledge & Kegan Paul, 1963) especially pp. 54–9, 215–25 and 240–8.
2. Salmon, *The Foundations of Scientific Inference* (University of Pittsburg Press, 1966) p. 125. Cf. G. Schlesinger's *Confirmation and Conformability* (Oxford: Clarendon Press, 1974) pp. 51–3.

3. Ibid., pp. 115–26.
4. See R. Purtill's *Logic* (New York: Harper & Row, 1979) pp. 316–23; I. Hacking's *Logic of Statistical Inference* (Cambridge University Press, 1965) pp. 190–207; and I. Levi's *Gambling with Truth* (New York: Knopf, 1967) pp. 139–56.
5. Salmon, *Scientific Inference*, pp. 124–32.
6. See Purtill's *Logic*, pp. 333–8; Hacking's *Logic of Statistical Inference*, pp. 208–27; and Levi's *Gambling with Truth*, pp. 139–56.
7. Cf. pp. 136–41 of M. Resnik's *Elementary Logic* (New York: McGraw-Hill, 1970).
8. Cf. L. Cohen's *The Implications of Induction* (London: Methuen, 1970) pp. 51–60.
9. Ibid.
10. The upshot of this, as it will be maintained in the next chapter, is that this kind of rapid confirmation requires that the hypothesis in question depicts an ontology *in common to all* of its rivals.
11. Schlesinger, *Confirmation and Confirmability*, pp. 101–5 and in his *Method in the Physical Sciences* (New York: Humanities Press, 1963).
12. I. Copi, *Introduction to Logic* (New York: Macmillan, 1978) pp. 486–92.
13. In fact, I think it may be the other way around: a standard of accuracy may acquire social-political value because of its epistemic utility.

7   Explanation and ontology

The time has come to review the various approaches to explanation and scientific understanding that have been covered in the text up to this point. What we are after, here, is an account of how it is that theories lead to rich and powerful explanations and deep scientific understanding of the occurrence of a variety of phenomena. This is what we have been searching for all along, namely, answers to such questions as how theories work to explain, predict, determine what we observe and so on. Our intellectual quest is essentially to develop a theory about the nature of theories. Various key formulations have been presented in Chapters 2 to 5 of the text, theories about theories that have played a dominant role in the traditional philosophy of science. According to the logical positivists, explanation and understanding were not all that complex. Once all theoretical language is translated into laws or correlations between observables, explaining a phenomenon amounts to demonstrating how it fits within this overall pattern of observables. Nature is understood in terms of how events systematically and economically repeat themselves. The D–N model, which renders explanation as the subsumption of a phenomenon under a lawlike proposition(s), is not really all that different from logical positivism, except that Hempel explicitly disavows the verification principle of meaning, allowing for the incorporation of theoretical entities in our explanans. As I have repeatedly stressed, this approach characterises explanation as an inference, deductive or inductive, from initial conditions to a description of the phenomenon to be explained; this takes place by means of laws that are either deterministic or stochastic. Hanson radically departs from the positivists and Hempel, abandoning the formal approach to theories and replacing it with a psychological, gestalt model. Understanding a phenomenon involves a gestalt, perceiving it as part of an overall pattern of events. Explanations bring this about by redescribing the phenomenon in question in higher-level, theory-laden terms. Explanations are semantical devices, the application of higher forms of concepts by means of which we are to organise phenomena. Hanson is a semantical contextualist, having the identity of a theory wrapped up in the content of the theoretical concepts it employs *vis-à-vis* its logical structure, the latter only being responsible for its inferential powers. Explanation and prediction are entirely different. Kuhn too, is a contextualist, one who is even more radically so than Hanson, for his contextualism goes all the way by including even the values and metaphysical beliefs of the scientific community as an essential part of the content of a theory. This is something that would be unthinkable for the positivists and formalists who cherish objectivity (and facts) as the essence of science. Explanation, according to Kuhn, is the assimilation of the phenomenon in question in terms of one's paradigm. This is accomplished by showing how a variety of phenomena fit in with the metaphysical beliefs of the scientific community, with the way the members of the community perceive nature, including their ability to employ techniques of calculation and prediction of phenomena that are within their values of accuracy.

It is my contention that while the logical positivists, Hempel and Hanson, each bring out something important and true about theories, they are mistaken about the source of a theory's explanatory power because of their misplaced emphasis on the language of a theoretical system. I think that a theory's ability to explain can be traced to what it says about the nature of things, how things work, i.e., the major explanatory work of a theory is carried out by its 'catalogue' of objects, their properties and the kind of causal interactions that take place among them. The view of explanation I am advocating is one very similar to Rom Harré's analysis, which can be found in *The Principles of Scientific Thinking, Matter and Method, Theories and Things*, etc. According to Harré, one of the major tasks of a theory is to present us with an ontological 'zoo' by means of which we are to comprehend events.[1] This is a version of ontological contextualism whereby the nature of an explanation is relativised to the kinds of entities, properties and interactions named by the theory, ontologies which vary from theory to theory, from the hard, localised Newtonian atoms – whose behaviour is controlled by forces – of classical mechanics to the four-dimensional geometrised space–time 'worms' of general relativity theory. Theories express a variety of ontologies, and if we really want to appreciate the powerful and often beautiful explanations they generate, we must go deeply into what they have to say about the nature of things. It is true that Hempel allows for ontological aspects in D–N but whereas he assigns ontology merely a 'pragmatic' role in explaining things, I maintain its role to be an essential one, even to the extent that how a theory predicts is logically subservient to the way it explains, i.e., to its ontology. Such a claim amounts to heresy for those who believe in D–N. Hanson, too, pays lip-service to the ontology of a theory but quickly focuses his attention on its semantical and psychological aspects. Even though things get better with the advent of Kuhn's metaphysical models, they are nevertheless just one element of a scientific paradigm among others, and they are not really given their due. He certainly never sheds any light on how metaphysical models explain phenomena – only that they are present in theory.

It is time, then, to give the highly neglected ontological aspects of a theory the credit they so richly deserve. The way I will do this is to complete the argument for the realist approach to theories which was outlined in the beginning of Chapter 6. In Chapter 6, it was argued that rapid confirmation of a theory (or an hypothesis) takes place to the extent that it can *account* for data or that results occur across categories. Now we want to know how it is that theories can account for or explain this kind of data. Data which occur across categories are a sub-class of what I will characterise below as *independent phenomena*. Premiss 3 of my argument for realism says that theories explain by reducing the number of independent phenomena and that this reduction takes place by identifying independent phenomena with aspects of an underlying common ontology. But if this is so, then what a theory has to say about the nature of things is not just a nicety or a pragmatic feature of psychological interest alone but it is essential if explanation and scientific understanding are to take place at all. This is not paying lip-service to the role theoretical entities play in a theory but an analysis which places complete emphasis on them. Premisses 2–4 of the argument and the conclusion in favour of realism are discussed in sections 2–5 where we finally utilise the results of Chapter 6.

1. A NEO-POSITIVISTIC APPROACH TO EXPLANATION

Recall the randomised force field case that was presented as a counterexample to the structural identity hypothesis in Chapter 3. The heart of this gedanken experiment was the irreducibly randomised force field which irregularly caused each bullet to be deflected on its way to the screen. And, if the reader will remember, this constitutes a direct attack on the notion of explanation being an inference on the basis of laws and initial conditions, for it was built into this case that it was, in principle, impossible to infer where each individual bullet would arrive at the screen; however, once each one completed its journey, we had a *post hoc* causal explanation available to us in terms of the field transferring certain amounts of momentum to each bullet. Such an account is warranted by the conservation of momentum. Even Hempel's inductive version of the structural identity hypothesis will not work. Explanation, here, would amount to showing or inferring that the bullet in question was very likely to end up at a particular place on the screen; but, in the above gedanken experiment, even though it may be very unlikely for a bullet to end up at a particular location on the screen, we can nevertheless explain how it ended up there in causal terms. This is just one reason, then, why we felt that explanations could not be identified with potential prediction.

Now, one neo-positivist solution of the above problem that has come into vogue quite recently is the Statistical–Relevance model ('S–R model') of explanation, whose sponsors have included Wesley Salmon, Richard Jeffrey and James Greeno.[2] The S–R approach to explanation rejects the view that explanations are arguments or inferences that the event in question had to occur or was likely to occur: 'An explanation is an *assembly of facts statistically relevant* to the explanandum, *regardless of the degree of probability* that results.'[3] Two nice things follow from this approach. In the first place, we no longer have a dualism of explanation corresponding to deductive and inductive reasoning; there is simply the assignment of statistical weights prior to an event's occurrence. Thus, D–N is simply a *limiting case* of S–R, where the probabilities assigned to events are either 1 or 0. Once free of the view that explanation is inference, what the proponents of S–R then do is boldly to assert that the assignment of an event's probability on the basis of its properties in relation to prior conditions suffices to explain why it occurred, *even if it is very unlikely that such an event would occur*.

The S–R theorists thus regard an explanation as 'a set of probability statements, qualified by certain provisos, plus a statement specifying the compartment to which the explanandum event belongs'.[4] Although such a tidy picture of a scientific theory may appear to many to be lacking in light of the seeming complexities found in theory, Salmon nevertheless makes this impassioned plea for S–R:

> When an explanation has been provided, we know exactly how to regard an *A* with respect to the property *B*. We know which ones to bet on, which to bet against, and at what odds. We know precisely what degree of expectation is rational. We know how to face uncertainty about an *A*'s being a *B* is the most reasonable, practical, and efficient way. We know every factor that is relevant to an *A* having property *B*. We know exactly the weight that should have been attached to the prediction that this *A* will be a *B*. We know all of the regularities (universal or statistical) that are relevant to our original question. What more could one ask of an explanation?[5]

One immediate reply to Salmon's rhetorical question might be: only an account as to *how B* takes place in relation to *A*! Interestingly enough, even though explanation is no longer an inference according to S–R, it is nevertheless in complete accord with the structural identity hypothesis, for it can be seen that a complete explanation yields a *basis* for making a prediction (and vice versa).[6]

On the assumption that S–R has taken care of the above problem of explaining the occurrence of a highly unlikely event, the real work ahead for the philosophers is to develop sophisticated probability calculii, including an interpretation of the fundamental concepts in these systems. While this approach to theories may be attractive to those of a positivist and formalist bent, I do not think that it can do adequate justice in capturing the explanatory power of theories. In the first place, it appears that explanations come much too easily with S–R, for we can find numerous cases in the sciences, especially in quantum mechanics, where we can assign probabilities to events in accordance with the principles of S–R while these events nevertheless remain a mystery; they are simply not understood. In other words, even though S–R appears to be a bold and logical solution to the problem of explanatory dualism which confronts D–N, it nevertheless is a model of explanation which defies or at least challenges our intuitions on understanding. S–R implies that we must accept this kind of an explanation of events: The reason why event, e, occurred was because it was very unlikely to occur. Presumably, the more accurate the probability of unlikelihood is, the better the explanans. Intuitively, such an account should leave one unsatisfied. Thus, a justification of S–R amounts, in part, to persuading one to overcome these intuitions. This is precisely what Salmon tries to do.[7]

Salmon's very own example of explaining the occurrence of an alpha particle penetrating a potential barrier in spite of the fact that it does not possess the required amount of energy to get through (which is picturesquely called the 'tunnelling effect') exactly illustrates this point; for even though quantum mechanics readily enables us to calculate the probability of such an occurrence (being an unlikely 1-in-10^{38} chance of making it through), we and the physicists still want to know *how the particle got through* or something like that. Accurately stating the unlikelihood of such an event does not answer *this* question. Here is his account of the tunnelling effect:

> To take a somewhat more dramatic example, each time an alpha particle bombards the potential barrier of a uranium nucleus, it has a chance of 10^{-38} of tunneling through and escaping from the nucleus; one can appreciate the magnitude of this number by noting that whereas the alpha particle bombards the potential barrier about 10^{21} times per second, the half-life of uranium is of the order of a billion years. For any given uranium atom, if we wait long enough, there is an overwhelmingly large probability that it will decay, but if we ask, "Why did the alpha particle tunnel out on this particular bombardment of the potential barrier?" The only answer is that in the homogeneous reference class of approaches to the barrier, it has a 1-in-10^{38} chance of getting through.[8]

Are we to be so accepting? As articulate as Salmon is about his proffered account of the tunnelling effect, we are still left with several unanswered questions which an understanding of so mysterious a phenomenon demands. For example, did the particle *break* through the potential barrier? If so, how could it if its energy was not sufficient to do so? (How could a wad of paper break a pane of glass?) Was there a 'rip' or a 'hole' in the barrier which permitted the particle to escape? Is the particle that ended up on the other side of the barrier the same as the one that was initially inside or was the original one destroyed and a new one created outside the barrier? And so on. I think that these are the kinds of questions that a scientist would naturally ask; yet, they are not answered by the above and, for these reasons, we still have a mystery on our hands. If the above account of the tunnelling effect can be construed as a standard case of an S–R explanation, then these nagging questions indicate that something is amiss with such a model.

There is something more fundamentally wrong with S–R, namely, it is open to be reduced to vacuousness. The basis of an S–R explanation, of course, is the assignment of a probability to a phenomenon B in relation to A. So it behooves us to ask how such probabilities are to be *interpreted*. Elsewhere, Salmon argues that probabilities are to be interpreted empirically in terms of the frequency of occurrence.[9] For example, to say that the probability of getting heads on a coin flip is equal to $1/2$ *means* that upon flipping the coin several times, the ratio of heads to tails will approach one. While such an interpretation of probabilities may help us 'keep our natural sciences empirical and objective', what happens to S–R explanations when the statistical part of S–R is given the frequency interpretation? Suppose, in the tunnelling effect case, we wanted to explain why a particular particle was found outside the barrier. According to S–R, the explanation would look like this: the reason why this atom was found outside of the barrier is because, given a potential field of this type and the laws of quantum mechanics, the odds are 1-in-10^{38} that such an event would take place. Now if we substitute the frequency interpretation of probabilities in the explanans of this propounded account: the reason why this atom was found outside of the barrier is because, in the past, when contained by a potential field of this type, 1-in-10^{38} of the atoms inside of the barrier got out. Now, is *that* why *this* particular atom got out? Unless our intuitions are completely awry, the S–R model of explanation, along with its adherence to the structural identity hypothesis, simply cannot allow us to find out what really happened in this case.

What Salmon has done is this: in the process of using the assignment of probabilities to explain events, no matter how unlikely they may be, he has equivocated on his explanandum. It is one thing to ask why an unlikely event occurred, where an appropriate response might very well be that even unlikely events occur once in a while if given enough opportunity. But it is quite another thing to ask *how* the event took place where an answer might include a description of some physical or causal process, even though such processes occur very infrequently (as was the case of the bullet landing at a very unlikely spot on the screen). This is partially why the tunnelling effect still remains a mystery even though quantum mechanics predicts its statistical probability. For this reason and others (which I will return to later on), I think it would be much better to handle the above problem of explaining unlikely events by throwing out the structural identity hypothesis altogether and assign completely separate logic roles to prediction and explanation within our theories.

While I strongly disagree with the S–R approach to explanation, I do think that it offers a positivist foil to the view of explanation I am about to present. Although Salmon has recently changed his mind about S–R – which will be covered at the end of section 4 – we have seen that the essential belief underlying S–R is that theories explain phenomena by subsuming them under probabilities for their occurrence. We also know that these probabilities are generated by induction. This is a beautiful case of reducing explanation to generalisations over the facts, which is quite in keeping with the positivist traditions.

A basic assumption of S–R, one which cannot be overemphasised, is that probability assignments to phenomena are provided by induction and these assignments serve as a basis or ground for explanation. However, contrary to Salmon, I contend that it is the other way around: *the probabilities that we ascribe to various phenomena are really a function of how theory explains their occurrence*. Theories which differ in the way they explain events will usually assign different probabilities to them or, at least, to combinations of them. In other words, we can not use probability assignments to characterise explanation because the former depends on the latter. This also means that how *a theory predicts depends on how it explains*, a reversal of the positivist contention that explanation is reduced to potential predictions. Likewise, since explanation is to be characterised in a way completely independent of prediction, the resultant analysis will be diametrically opposed to Hempel's structural identity hypothesis.

In the next section, I argue that theories explain by showing how phenomena which seem, at first, to be physically unrelated are actually aspects of a common or underlying ontology; that is, *theories explain in virtue of their reference to objects*. Theories that are committed to different types of entities – different ontologies, such as atoms, waves, fields, etc. – then explain phenomena differently and, hence, assign different probabilities to combinations of events. (There are exceptions to this, namely, where two different theories assign the same probabilities to events, but we need not be concerned with such a situation here.) The idea put basically is this, and it should not be so surprising: theories that have different things to say about the nature of things almost always yield different probability assignments for events. This is important, for once this connection between the ontological commitments of a theory and its assignment of probabilities is established, the advantage of the realist approach to theories over that of the logical positivist is made clear. This will be carried out in section 5.

2. EXPLAINING BY REDUCING THE NUMBER OF INDEPENDENT PHENOMENA[10]

The type of explanation I wish to focus upon here is as old as Thales's attempt to account for everything in terms of water and as new as the use of quark theory to incorporate a variety of microparticles and forces. In *Probability and Induction*, William Kneale comes up with an original and intriguing idea on the explanatory power of theories. Kneale maintains that a theory's explanatory power can be traced to its ability to 'reduce the number of independent phenomena'. According to this view, explanations succeed in 'reducing the number of independent laws we need to assume for a complete description of nature'. For example, Newton's laws explain the phenomena covered by the Boyle–Charles law, Graham's law, Galileo's law of freefall because the latter, independently established laws can be derived from Newton's laws alone. In this way, the laws of diverse science which were previously thought to be unrelated are now united in a single framework.[11]

Recently, Michael Friedman has further developed this idea and attempts to make Kneale's intuition on explanation more precise by assuming that phenomena can be represented by (lawlike) sentences. He often proposes a definition of a sentence explaining a set of sentences in terms of a formal definition based on the properties of (lawlike) sentences. The basic idea here is that before Newton's laws, the above array of laws and their consequences were thought to be independently acceptable. However, once we realise that they can *all be derived from Newton's laws*, the number of independently acceptable consequences is drastically reduced.[12] As much as I favour Friedman's (and Kneale's) major intuition – that we explain and increase our understanding by reducing the number of independent phenomena – I feel that, if anything, Friedman's formal characterisation of this idea interferes with truly appreciating the notion of reducing independent phenomena.

Not only have fairly insurmountable, technical objections taken Friedman's formal definition to task,[13] I think it is a mistake to assume that the reduction of independent phenomena is accomplished by the derivational properties of lawlike sentences in the first place. For one thing, it is the number of independent *phenomena* that is reduced, not the number of laws. It is entirely possible for such a reduction to take place where the number of independently acceptable laws has increased! In order to show this possibility, let us consider this imaginary case. Suppose that instead of what is currently believed about the laws of physics, there are many different elementary particles which are characterised by having various amount of 10 000 different charges, i.e., there are, for the sake of exaggeration, 10 000 different forces which are characterised by at least 10 000 different and independent lawlike statements. In terms of these elementary particles and their multifarious interactions, we can give an account of a vast array of macroscopic phenomena – e.g., radioactive decay, gas behaviour, the refraction of light, even gravitation, etc. Here we have a number of independent phenomena which can be viewed in terms of a single theory of elementary particles and their interactions. But, intuitively, this takes place *not* by somehow reducing the *number* of relatively independent lawlike sentences since we, in effect, have added at least 10 000 new, independent ones.

Just as important, because a set of laws may be reduced and explained, the above characterisation of explanation makes it extremely difficult if not impossible to explain individual events. After all, phenomena are not just laws but events, the behaviour of properties and individual entities, as well as the behaviour of collections of things. If events are construed as independent phenomena, how is it that *they* are reduced? It can not be that the number of laws that cover them is pared down. Suppose events e_1 and e_2 are covered by law_1 and e_3 and e_4 are subsumed under law_2. If we were to subsequently derive law_1 and law_2 from law_3, what sense would there be in saying the number of independent events has been reduced, except to say that e_1–e_4 are now subsumed under a single law? However, e_1–e_4 can still be construed as independently occurring events. I feel that there is more to reducing the number of independent phenomena when it comes to events than simply to say that they are covered by fewer laws. What this sense of event reduction is will be delineated below.

Can the Kneale–Friedman intuition be saved? I think it can, and in such a way as to give us some insight into how the ontology of a theory generates explanations; in fact, the reason why I believe their formulation is open to these counterexamples and why they are unable to incorporate explaining events is because 'dependent' and 'independent' phenomena are defined in logical or derivational terms instead of ontological terms.

There are two, related senses of reducing the number of independent phenomena. The first sense concerns their existence, i.e., two phenomena are independent in that there is no lawlike connection whatsoever between the existence of one and the other. Let us look at a case where independent phenomena are events or occurrences to a particular individual or object. Suppose, for example, that an individual is plagued with a chronic cough and a low-grade fever. On the surface of things, the existence of each phenomenon has nothing to do with the other. Not so, when they are viewed as the symptoms of a disease such as bronchitis. While the inexperienced victim may not see any connection between them, his physician, who is quite familiar with the workings of the disease, realises full well that it is medically impossible for one of these symptoms to occur without the other. And this is the first sense of 'independent' phenomena I have in mind when we speak of a theory reducing the number of independent phenomena: *two phenomena are independent in that it is* (physically) *possible for one to occur without the other and vice versa.* The possibility or impossibility of events, or state of affairs, here, is determined by the laws of nature. Something is possible if it is in accordance with the laws of science while it is impossible if it violates any of these laws. For example, it may be chemically possible to form a water molecule with three hydrogen atoms (H_3O) but impossible to form one with five hydrogen atoms (H_5O) because the former is in accordance with the laws of chemistry while the latter is not. (This 'combinatorialist' sense of possibility will be detailed in Chapter 9.)

Within this first sense of 'independent', there are two ways to speak of reduction. The first concerns the occurrence of events under a particular set of conditions or circumstances. In this case, independent phenomena are reduced and explained by showing how, under these specific conditions, it is physically impossible for one event to occur without the other and vice versa. Thus, while the unwitting disease victim could really conceive of not having a low-grade fever accompany his cough, his doctor could not, because he knew that his patient had a disease of which the two are symptoms. The second type of reduction in the first sense of 'independent' concerns the *occurrence of different kinds of entities in nature* such as gases, liquids, solids, etc. They can be considered independent phenomena in that their coexistence could very well be physically accidental. It is quite conceivable, for example, for gases to exist but not solids and the reverse. Instead of talking about the occurrence of individual events under a single set of conditions, we now speak of the existence of one type of entity under specified conditions, such as the occurrence of water at a given pressure and temperature, and another type of entity, such as a gas, under other circumstances. Reduction, here, amounts to showing how the conditional existence of one type of entity is linked up with the conditional existence of another type. For example, kinetic theory tells us that it is physically impossible for gases to exist under one set of conditions but solids not to exist under another.

There is another but related sense in which it can be said that phenomena are independent. It occurs just in case the properties (and behaviour) of one are independent of those of the other, i.e., the value of any property of one is functionally independent of the value of any property of the other. Another way to put functional independence is to say that there are no lawlike relationships holding between any of the properties of one phenomenon and those of the other. So while there are laws that cover the behaviour of gases and solids there are no laws (before reducing these independent phenomena) that go *across* these systems. For another case, before the advent of Newtonian mechanics, Galileo's law of freefall and the law of parabolic motion for projectiles covered terrestrial phenomena while Kepler's laws strictly applied to the extraterrestrial motion of the plants; but there were no lawlike relationships holding between the values of the variables which described these two diverse phenomena. We shall learn that one result of adopting a Newtonian ontology is that there exist lawlike relationships holding between variables which describe terrestrial and extraterrestrial phenomena. Thus, these independent phenomena are reduced in the second sense.

How, then, do theories bring about the reduction of the number of independent phenomena in the above two senses of 'independent'? My answer is that theories reduce the number of independent phenomena and thereby explain them *by providing them with a common ontology*. Isn't this why atomism was such an attractive theory, even before it led to powerful predictions? For a vast array of seemingly unrelated phenomena, it provides a common ontology of atomic configurations and motion. In the same way, the mass points and forces of Newtonian mechanics served to unify terrestrial dynamics and extraterrestrial behaviour of the planets.

We can make the above notion of reduction more precise: theories that explain by reducing the number of independent phenomena do so *by showing that these seemingly separate and diverse phenomena are all manifestations of the same system of objects*.

Different phenomena, $P_1, P_2, P_3, \ldots, P_n$ are accounted for in terms of a common ontology by *identifying* $P_1, P_2, P_3, \ldots, P_n$ with different aspects of that ontology.

$P_1, P_2, P_3, \ldots, P_n$ are identified with different aspects of the same set of objects, processes, etc., in the same way that a set of six squares may be identified with six surfaces of a cube. What we have here, semantically, are two definite descriptions which are connected by an identity relation, where one of the descriptions takes on the form: '*aspect of*____', '*manifestation of*____', '*appearance of*____', and so on. These expressions are what J. L. Austin rightly calls 'substantive-hungry' because the noun before the preposition requires completion by reference to an entity or process.[14] In this case. our independent phenomenon is seeking some part or aspect of a common ontology with which to be identified while the remaining independent phenomena will perhaps latch on to other aspects.

A few remarks about the notion of a common ontology. In the above-mentioned case of several types of imaginary particles and the forces which govern their behaviour, what could possibly prompt us to say that these 10 000 different particles comprise a common ontology? The answer is that although they all differ from one another in terms of properties and behaviour, they nevertheless have *properties in common* such as charge, spin, momentum and energy. In fact, they should have properties in common to the extent that laws can be formulated to handle all possible combinations of interactions among different particles. (I can not argue for this here but I suspect it can be shown that such properties in common are necessary if we are to have laws which cover combinations of interactions.) So a common ontology, at the least, consists of a set of objects which share properties to an extent that any combination of many-bodied systems is covered by laws.

There is a crucial reason for having the relationship between independent phenomena and aspects of a common ontology be that of identity. The identity relationship, '$a = b$', says that a and b are one and the same thing. Once two things are established as being the same, a very powerful logical or semantical principle comes into play, known as Leibniz's *indiscernibility of identicals*: if two things, a and b, are the same, anything that is ascribed to a must be ascribed to b and vice versa. For example, supposing that Superman and mild-mannered Clark Kent are one and the same individual, if Superman weighs two million pounds (because of his extreme density) so must Clark Kent. If Superman has the capacity to leap buildings in a single bound, so does Clark Kent. If Superman can be affected by Kryptonite so can Clark Kent. If Clark Kent is mortal so is Superman. The mutual transference of ascription would follow for any property or relation, whether we realise it or not, given that the two are actually the same. It is very important for the reader to understand this principle and be able to apply it, for it will be utilised constantly throughout this chapter and the next. The semantics of the identity relationship will be stated in more precise terms in Chapter 8. But for now, I will note one implication of this principle when it comes to individuals and lawlike relations. If $a = b$ and L is a lawlike relationship which holds for a then L holds for b (and vice versa). Put another way: if $a = b$, then they must behave in exactly the same way. So we see that if a particular independent phenomenon is *identified* with an aspect(s) of some set of objects, the laws which hold for that set of objects now apply to the independent phenomenon and any laws pertaining to the independent phenomenon must also hold for those things with which it is identified. (The reader should be warned of the extensive literature on the problems of applying Leibniz's principle in various contexts. A discussion of these issues, however, requires a very advanced background in mathematical logic and the philosophy of language.)

Let us now examine how identifying $P_1, P_2, P_3, \ldots, P_n$ with aspects of a common ontology provided by theory reduces the number of independent phenomena in the above two senses of 'independent'. In the case of reducing the number of independent *events*, I am assuming that we can sensibly speak of event identity, i.e., we can speak of two events being the same or different.[15] Suppose, then, that the originally construed independent events are now each identified with events that take place among the objects which comprise the common ontology provided by theory. Events that occur to objects of a common ontology do so in a lawlike way: by Leibniz's indiscernibility of identicals, then, we realise that the laws which apply to objects of the ontology also apply to events taking place among independent phenomena. As a result, it is impossible according to the laws of nature for one of the originally construed independent events to exist but not the others. They 'cease' to be independent phenomena. In the above medical example, if the cough and fever are identified with various stages of an underlying biological mechanism, the laws which control the course of the mechanism also cover the occurrences of the cough and fever. Thus, it is no more possible for one to occur without the other, in the same way that one aspect of the biological mechanism cannot possibly take place without the others.

According to kinetic theory, gases, liquids and solids, etc., are all manifestations of atoms under specified conditions. Even though we can conceive of the universe consisting of nothing but gases, we know by atomic theory that this is tantamount to saying that the universe exists under highly specified conditions and no others (say, the temperature everywhere the atoms are located is one million degrees Kelvin). If, on the other hand, a greater variety of conditions existed, atoms would take on other forms, as liquids and solids, for example. In other words, the laws of our common ontology connect the nature of atomic structure under some circumstances with all others. Now kinetic theory identifies gases, liquids, solids, etc. with aspects of atomic configurations under a variety of conditions. Assuming that the laws of nature express some sort of physical necessity, it is then physically impossible or inconceivable for a gas to exist under one set of conditions and a solid not to exist under another set. In the same way, the laws of atomic physics tell us that it is physically impossible for water molecules to move about freely at a temperature of one hundred degrees centigrade or greater but not form an atomic lattice at a temperature of zero degrees centigrade or below. Thus, one result of identifying independent phenomena with aspects of a common ontology is that the existence of various kinds of entities in nature becomes connected.

The second sense of reducing independent phenomena naturally follows. We might believe that the property denoted by the temperature of a gas may have no relationship at all with the rigidity of a solid or that the rate of fall of a body released above the Earth is not functionally related to the period of the moon, but once these features are identified with aspects of a common ontology, their functional dependence ensues. The gas and solid case goes roughly like this. Once gases and solids are viewed as atomic configurations, atomic theory tells us that if we lower the temperature of, say, water vapour it will condense into a fluid and that if we lower it even further, the attractive forces among the water molecules will overcome their kinetic energy, forming a rigid lattice. So, the rigidity of ice is partly a function of temperature but the temperature of a solid is the very same functional property denoted by temperature when applied to a gas. In the case of the falling body and the orbit of the moon, there was no apparent (lawlike) connection between the values of the variables in the law of freefall and Kepler's laws before Newton's physics. Yet, one can readily be found upon reducing these independent phenomena: the time a body falls towards the Earth in $S_1/S_2 = t_1{}^2/t_2{}^2$ (law of freefall where 'S' denotes the distance fallen in time, t) is the same variable in Kepler's third law when it is applied to the moon's orbit ($A_1/A_2 = t_1/t_2$, i.e., the moon sweeps out equal areas in equal times). Thus, if falling apples (a terrestrial phenomenon) were suddenly to orbit the Earth (an extraterrestrial phenomenon), the area swept out by each apple would be functionally related to how far it falls at each increment of time.

Let us consider another, rather picturesque example which illustrates the two ways of explaining by reducing the number of independent phenomena. A nice thing about this example is that only high school geometry is required to appreciate the intricacies of how a common ontology works to explain events. It comes from Edwin Abbott's *Flatland*,[16] which is a text designed to help the student and layman understand what it would be like to use four-dimensional geometry to explain three-dimensional events. Imagine a person visiting a two-dimensional creature who lives in a place called Flatland. During the course of this friendly get-together, they observe the following sequence of events: a point appears before them and immediately becomes a circle which increases gradually to a maximum diameter and then, just as gradually, diminishes to a point. This sequence of events leaves the Flatland creature completely perplexed. Not so his three-dimensional visitor, who offers this account of what just happened. 'These events', he says, 'are not so mysterious as they may seem to be. Afterall, what you just saw was a sphere passing through a plane.' Now if this perfectly reasonable explanans is to make any sense to the Flatland creature, in spite of the fact that he is a two-dimensional creature who can not possibly observe anything three-dimensional, he must have some idea what a sphere is, what it is like for a sphere to move in a three-dimensional space, the result of a sphere intersecting a plane (i.e., a conic section), and so on. This knowledge is essentially about the ontology of a higher-dimensional world. The independent phenomena, here, are events which consist of two points and a series of circles appearing in Flatland. Viewed as circles and points, it appears to be quite possible, under these circumstances, for any one to occur in isolation (independently) from the others. Likewise, there appears to be no reason to believe that there exists a functional relationship between the radius of any one circle and that of the others. As things stand, the ordering of the circles in terms of their size *could have been otherwise*, i.e., if we are just dealing with circle events, we could have a large circle followed by a point and then a small circle followed by a large circle. Any combination of events would have been possible. So, the occurrence of points and circles in Flatland certainly appears to be independent in terms of existence and functional relations.

'What you just saw was a sphere passing through a plane' provides understanding by eliminating independence among these seemingly isolated events. Each circle event is identified with the occurrence of a conic section (= the intersection *of* a sphere and a plane). As a result, so long as the sphere passes through the plane – given, at a constant velocity – no one circle could possibly exist when and where it does without the others unless the laws of geometry are violated. Their independent existence is a geometrical impossibility. For the same reason, the diameter of each circle must be a function of the diameters of all the others, and we now have a lawlike relationship going across independent phenomena, i.e., it is geometrically impossible for the value of the diameter of any one circle to be independent of that of any other in the sequence. This relationship depends on the curvature of the three-dimensional object. The greater the curvature the faster the rate of increase (and decrease) of the diameters. An ellipsoid will yield a slower rate of increase and decrease of diameters because its curvature is less than that of a sphere. So, even if the Flatland creature cannot directly observe a sphere or an ellipsoid, the functional relationship among the circles would enable him to distinguish between the two. Again, this reduction of independence takes place because each circle is *identified* with the intersection of a sphere and a plane in the same way that a square might be identified with a particular surface of a cube. I emphasise that the circles and conic sections are neither causally related nor correlated but are simply one and the same. This will be of paramount importance when the type of explanation we have just explored is compared to Hempel's and Salmon's approach. We shall learn that the identity relation in conjunction with the indiscernibility of identicals is the key to such a comparison.

Getting back to reality, a most recent case of explaining by reducing the number of independent phenomena can be found in particle physics. As it was briefly mentioned in Chapter 3, the new physics interprets forces in terms of the transference of messenger particles. For example, the deflection of two electrons is seen as the result of the transference of a photon which is emitted by one of the electrons and is, in turn, absorbed by the other. Now one of the many, deep mysteries that has cropped up in the study of the atom concerns the fact that protons (positively charged particles of a mass, 10^{-22} grams) and neutrons (which have the same mass as the proton but no electrical charge) are bound in the atomic nucleus. This is a mystery because that they are so bound cannot be accounted for in terms of either gravitational or electrical attraction. (The former force is much too weak while the latter does not operate on neutrons.) All that could be said was that there existed a powerful nuclear force which was responsible for binding particles in the nucleus, without the slightest idea as to how such a force operated. The first clue came in the 1930s and 1940s when short-lived force-carrying particles, called mesons, were postulated and subsequently discovered as being responsible for nuclear binding. These particles kept protons and neutrons trapped in the nucleus by zipping back and forth among them. This somewhat fuzzy picture of nuclear attraction came into sharper focus with the advent of quark theory. Quark theory fundamentally revised the ontological zoo of particle physics. According to this new particle picture, the hitherto previously thought basic, irreducible proton and neutron turn out to consist of triplets of quarks, each quark with a 1/2 spin and whose amount of electrical charge varies according to its spin. The proton and neutron, then, are not really fundamentally different particles, as was once believed, but are instead simply *different aspects of the quark ontology*: the proton consists of three quarks that add up to a plus one charge ($2/3, 2/3, -1/3$), while the neutron is made up of three quarks whose charges ($2/3, -1/3, -1/3$) cancel out to zero.[17]

By identifying a variety of particles with combinations of quarks, many new accounts of their behaviour came into being. In particular, the nuclear force which bound the proton and neutron in the atomic nucleus was now interpreted as the exchange of a 'quarky meson' between a spin-up quark in the proton and a spin-down quark in the neutron (as shown in Figure 7.1).[18]

So, in answer to the question, 'How is it that neutrons and protons are bound in the atomic nucleus?': protons and neutrons are kept together within the nucleus by a quarky meson which is

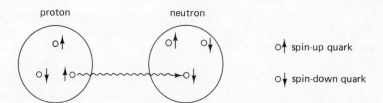

FIGURE 7.1 *Transference of a meson between two quarks*

exchanged between individual quarks in each proton and neutron.

The above theoretical account of nuclear binding is another way of reducing independent phenomena by providing a common ontology. The independent phenomena of our explanandum are the locations of individual protons and neutrons. (Imagine separate photographs of individual protons and neutrons, their composite forming a picture of a nucleus.) Each proton and neutron is identified with a combination of spinning quarks. Once this is done, their locations are no longer independent phenomena, their togetherness being the result of the above described exchange of a quarky meson among the quarks within each proton and neutron.

Another beautiful and most exciting result of the innovation of a quark ontology is that it provided a systematic way conceptually to handle the highly unexpected variety of particles that were discovered by the atomic physicists. (Fermi was noted to have said: 'If I could remember the names of all these particles I would have become a botanist.') Now they are united as combinations of quarks, the price being that they are no longer fundamental. This result may also hold for the variety of forces found to exist in nature. The scientific reasoning involved is just one more case of thinking in terms of reducing the number of independent phenomena. The strategy is to identify the variety of forces which are found in nature (gravitational, electrical, magnetic, weak and nuclear forces) with combinations of exchanges of messenger particles. Admittedly, the monumental task of formulating a unified theory of forces and particles is presently at a highly embryonic stage of development; however, research done in the area of physical symmetries has led to speculation of the existence of an even more fundamental 'superparticle' of which all others are combinations and an even more fundamental force process called supergravity:

Extended supergravity can also be viewed in terms of a "superparticle" with an arrow [which represents its state] in an auxiliary space of many dimensions. As the arrow rotates, the particle becomes in turn a graviton, a gravitino, a photon, a quark and so on. The quanta [particles which are exchanged in the process of forces at work] of all the forces are present in the theory, and they are unified, or derived from a common source.[19]

If such a theory of supergravity and superparticles were ever to come to fruition, realising Einstein's dream to unify the once-fundamental gravitational and electromagnetic force fields, it would lead to an ontology which would reduce the number of independent phenomena found in nature in the most intellectually glorious manner ever envisaged by the scientific world! Once more, this is done by identifying the variety of particles found in nature with aspects of a common ontology provided by a deeper and more unified theory, i.e., with combinations of superparticles and their states while nature's forces are identified with combinations of a single virtual-particle-exchange process.

While explaining by reducing the number of independent phenomena is a major type of explanation to be found within the sciences, it is not at all clear if all other kinds of explanations can be reduced to this type. Whether a commonsensical, everyday explanation such as 'Johnny went to the grocer's because he wanted bananas' is actually a case of reducing phenomena is quite an open question and it cannot be covered here. I really do not know to what extent explanations given in ordinary, everyday situations presuppose a common, underlying ontology. But the ancient people who may have explained the sudden darkening of the moon by saying that the gods were angry and swallowed it (whereas we speak of an eclipse) are presupposing some kind of an ontology, albeit one which differs from ours. And they have supplied *an* account of the phenomenon although we would find such an explanation unacceptable. Nevertheless the above analysis need not be extended to all types of explanation in order to be useful in capturing a major class of them. Perhaps it is too much to ask of any model of explanation significantly to cover all possible types. It is only intended to cover those cases where phenomena are explained by conferring some kind of unity among them. Just as important, this explanation is particularly well suited for understanding the existence and relations among a vast array of radically diverse phenomena, and it is precisely this kind of phenomena which will be used to compare the logical positivist and realist approaches to theories. Finally, it turns out that the type of explanation we have been examining here differs in logic with respect to Hempel's D–N model. I will now turn to examining this feature.

3. THE LOGICS OF EXPLANATION AND PREDICTION

Explaining by reducing the number of independent phenomena does have a definite logic but one which differs from that described by the D–N model. Once this is brought out, we can proceed to place the last nails in the coffin of the structural identity hypothesis. The basic idea is to show that two distinct inferences take place when it comes to explanation and prediction. Remember that according to D–N and the structural identity hypothesis, there should be just one.

Now, on the basis of the groundwork laid out in sections 1 and 2 of this chapter, we can readily distinguish between explanatory and predictive reasoning. In the case of explanation, we start out with independent phenomena and then identify them with aspects of a common ontology. But, in doing so, *no predictions are made*. Even if the cough and fever are identified with various biological features of the disease, this alone does not tell us when the next coughing attack should occur. On the other hand, once we know how it is that independent phenomena are identified with aspects of a common ontology, *if* we know the probability of the occurrence of phenomena pertaining to the common ontology, we can then go back and predict the occurrence of the independent phenomena. But fixing probabilities for the occurrence of common ontology events and then, on the basis of this, going on to infer probabilities for the occurrence of independent phenomena, is an entirely different inference. It is one thing to identify independent phenomena with various aspects of a common ontology but quite another to predict their occurrence on the basis of events which take place among the entities that comprise the common ontology. Returning to the Flatland example, we explain the occurrence of a sequence of circles in Flatland by identifying them with conic sections produced by a sphere passing through a plane. Suppose, on the other hand, we wanted to predict – *on the basis of our theory vis-à-vis* an inductive prediction which is based on past experience – if this sequence of circles will occur in Flatland again. Assuming that this particular type of sequence of circles is identified with a sphere passing through a plane, the probability of their occurrence, of course, depends on the probability of the next sphere's journey through Flatland. Again, this inference is not the same inference as identifying circles with intersections of a sphere and a plane. Predicting involves the assignment of probabilities to events but identifying independent phenomena with an underlying set of objects does not even deal with probabilities.

We must also be careful not to confuse the relationship between independent phenomena and their common ontology with the logical relationship between an explanans and explanandum. The former, of course, is not a logical relationship in that things are, in a sense, physically, not logically, related. The relationship between independent phenomena and their common ontology is more like that of parts to the whole. In the Flatland case, each circle is just one part or aspect of a sphere, the sphere and the circles being topologically related in a special way. But explanations are not things but sentences and propositions which are linked up in a certain logical way to the representation of the explanandum phenomenon.

Given that we explain by reducing the number of independent phenomena, we can represent the relationship between the explanans and explanandum. In order to show this, however, I will have to introduce the reader to a few basic set theoretical notions. In the first place, we can use sets to represent the explanans (common ontology provided by the theory) and the explanandum (independent phenomena). Each element in the explanans set will stand for a particular aspect of a common ontology while each element in the explanandum set represents an independent phenomenon. In the Flatland case, one set would include elements which stand for different conic sections while the elements of the other set would represent the occurrence of circles in Flatland. According to this way of viewing things, the problem of capturing the relationship between the explanans and explanandum becomes a matter of how the explanans and explanandum *sets* are related. It is my contention that we can most adequately relate them by what is known in set theoretical parlance as a mapping function. Consider two sets, A and B. To say that there exists a function or a mapping from A to B means, intuitively, that there is a rule, F, that associates one and only one element of B with each element of A.

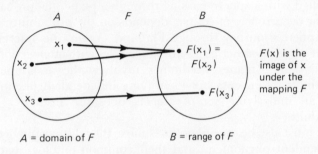

FIGURE 7.2

So the relationship between the set which represents independent phenomena, P_1, P_2, ..., P_n, and the set representing their common ontology is that of a mapping, F, with the former as the domain of F and the latter as its range. The elements of the domain and range of F are related by identity which is of the form of the above substantive-hungry expressions, P_n = an aspect ——————— (the blank being filled in accordance with our theory's ontological zoo). For example, in the Flatland case, the occurrence of each circle would be represented by an element of A; this element would hit an element in B which corresponds to a conic section of a sphere and a plane at a particular stage in the process of a sphere passing through a plane. (Even though each element in the domain of F is related to its image by identity, we want to allow for the possibility of *different* independent phenomena being identified with one and the same aspect of a common ontology. Thus, F is a many–one mapping.)

There is another important feature of F which results from the fact that independent phenomena are identified with various features of an underlying ontology. Remember how the identity relation, $a = b$, implies via Leibniz's law that any relationship, lawlike or otherwise, which holds for a must hold for b and vice versa. So, any relationship holding among independent phenomena must show up among the corresponding aspects of the common ontology. In the Flatland case, if the area of each circle is related to its radius, by πr^2, then that same relationship must hold for each conic section. If the radii of the circles are related by a particular function of time, the radii of the conic sections are related by the very same function. Set theoretically lawlike relations are represented as operations on elements of the set. This means that any operation among the elements in the domain will be preserved as will go to the range of *F*. This represents the preservation of laws as we go from independent phenomena to aspects of a common ontology. We are skipping over many technicalities, here, but a mapping where these relations are so preserved as we go from domain to range, is known as a homomorphism, *H*. So, the relationship between the explanans and explanandum is a homomorphism, *H*, i.e., a many–one mapping where any operations among the elements in the domain also appear in the corresponding elements in the range. Such a mapping mirrors the way independent phenomena are related to their underlying field of objects.

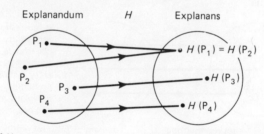

Explanandum   *H*   Explanans

$H(P_1) = H(P_2)$

$H(P_3)$

$H(P_4)$

domain of *H* = set representing independent phenomena $P_1, P_2, ..., P_n$ range of *H* = set representing aspects of a common ontology

FIGURE 7.3

The first thing to notice right away about *H* is that if *H* captures the relationship between the explanans and the explanandum, and if *H* can be thought of as being an inference at all, the direction of inference is *opposite* that described by the D–N model, since the direction of inference goes *from* the data or independent phenomena *to* the common ontology which is used to explain.

What, then, about prediction? Contrary to what the logical positivists and adherents of the structural identity hypothesis have attempted to teach us all along, a theory's explanatory power is actually responsible for the prediction base. *H* is a map which represents how a theory's ontology reduces the number of independent phenomena. But we know that *H* does not make predictions. However, on the basis of a mathematical theorem, we can derive the following relationship: the map H from the set of the independent phenomena to the set representing their common ontology induces or generates *another* map, H^*, this time, *from* the probabilities of the latter *to* the former. *It is this induced map, H^*, which serves as a theory's basis of prediction.* In other words, once we learn how a theory reduces the number of independent phenomena, which amounts to determining H, we can then determine another mapping *from* the probabilities of the subsets of the theoretical or explanatory system *to* the probabilities of the subsets of the explanandum system. So, H^* tells how to predict the behaviour of the independent phenomena on the basis of the probabilities assigned to their common ontology. The direction of H^* is, again, *opposite* that of H but in keeping with D–N. So the proponents of the structural identity hypothesis have essentially failed to distinguish H from H^* recognising H^*, only, as the logic of a theory.

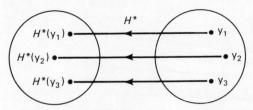

y_n corresponds to the probability for the occurrence of a particular ontological aspect while $H^*(y_n)$ stands for the probability that an independent phenomenon will occur

range of H^* = set representing probabilities of combinations of P_1, P_2,..., P_n

domain of H^* = set representing probabilities of combinations of aspects of a common ontology

FIGURE 7.4

The consequences of these mappings which correspond to explanation (H) and prediction (H^*) are all too numerous to mention. I will consider what I take to be just a few of the most important ones. We have seen that the logical structure of a theory is much too complicated to allow for the identification of the logic of explanation and prediction as the D–N model entails. We now realise that a theory has *two* logic or inference schemata, and unless D–N proponents can somehow conflate H and H^*, we can say farewell to the structural identity hypothesis. Once we free ourselves of the structural identity hypothesis, it is just a simple matter to avoid the dualism of the logic of explanation which followed immediately from Hempel's idea of explanation as an inference. We no longer have two kinds of explanations, corresponding to deductive and inductive bases of prediction. Now there is just one way we reduce the number of independent phenomena with just one logic of explanation which is represented by a single mapping, H. On the other hand, the induced mapping, H^*, serves as a theory's prediction base in that it is a mapping from the probabilities of events that take place within the common ontology (provided by the theory) back up to the probabilities for the occurrence of independent phenomena. And here, we can readily have two kinds of inferences corresponding to statistical and deductive reasoning: statistically-based predictions are those where the prior probabilities over our common ontology are between zero and one while, of course, a deductive prediction base means that these probabilities are either zero or one exclusively. So it all depends on how the theory fixes prior probabilities for events. How this is done does not affect how independent phenomena are reduced (or H). It is the other way around! I will refer to these explanatory and predictive schemata as H-H^*.

We have seen that one major contrast with D–N and the above H-H^* model is that while D–N considers prediction and explanation as one and the same inference from laws and initial conditions, H-H^* splits them in two. Since prediction, H^*, is induced from H, it turns out that how a theory predicts depends on how it explains, i.e., *how* it reduces the number of independent phenomena. Theories which depict nature differently vary in how they reduce independent phenomena and, hence, differ in the way they predict combinations of events. This point will be taken up in detail in the next section. But there are other, salient differences between D–N and H-H^*. This second one concerns the way in which each explains. The reader knows that D–N regards explanation as an inference from laws and initial conditions. According to H-H^*, explanation is not really an inference at all but unity and understanding are produced, instead, by *reference*, that is, by referring independent phenomena to various facets of an underlying nature. Theoretical entities, then, do not directly play a deductive role in explanation; their actual role is that of objects of reference, whose different aspects surface as independent phenomena. By this means, the existence, properties and behaviour of the latter are redescribed in terms of the former.

Thirdly, we have seen that H-H^* explains by eliminating contingency among events but it does so not by deducing their occurrence from laws and initial conditions. We have this situation instead. First of all, there is a series of phenomena whose existence and properties are contingently related, i.e., it is possible for any one to exist and possess the properties it has without the others. We identify phenomena with various features of a common ontology. The objects of this ontology obey certain laws, i.e., certain combinations of events and properties must occur the way they do in accordance with these laws. By Leibniz's indiscernibility of identicals, the necessity is transferred to the original but no longer independent phenomena. It is not empirical laws but Leibniz's principle and the identity relation which is primarily responsible for generating understanding and the elimination of contingency.

This leads to the fourth and final comparison between D–N and $H–H^*$. It concerns the role of laws in these two systems. We know that, for Hempel, laws were the essence of explanation as the major vehicle of the deduction and prediction of events. Laws have no such place in $H–H^*$, partly because explanation is not an inference in $H–H^*$. This does not mean, however, that laws do not have importance in $H–H^*$. We have just seen that we must always keep track of them in order to identify independent phenomena with combinations of theoretical entities without violating the indiscernibility of identicals. This leads to a crucial role of laws in explanation although it is not that of D–N. Before Newton's laws, atomism was just a programme, telling us to identify water, earth, fire, air, souls, etc. with configurations of atoms; but which ones? Atomism, at this stage, does not answer this question. A precise explanation requires that we pinpoint our identities, that we have a means of deciding which independent phenomena are identified with what aspects of common ontology. For example, if gases are identified with configurations of atoms, which configurations? The answer is: those configurations of atoms *whose lawlike behaviour is exactly the same as that of gases*. This, then, is where laws come into $H–H^*$. By their means, alone, we are able to determine the identity function in H. So, instead of their role being the key to deducing that phenomena will occur, as it is in D–N, the main function of laws in $H–H^*$ is epistemological in nature. *They are reference guides.* Of independent phenomena and aspects of a common ontology, they tell us how it is that the former are to be identified with the latter.

H–H* and D–N will be further compared with respect to the reduction of theories in the next chapter. For now, I will return to the above-mentioned dependence of *H** on *H* and see what lessons it may have for the S–R approach to explanation.

4. ONTOLOGY AND PREDICTION

We are now in a position to compare Salmon's S–R with *H–H**. According to S–R, explanation is simply a matter of systematically assigning probabilities to events. Salmon says that this is done in accordance with the frequency interpretation of probabilities, by observing the ratio of positive instances of the phenomenon in question (given certain initial conditions) to the total sample (over all possible phenomena). Again, such a view is well within the empiricist–anti-metaphysical tradition, where the role of ontology in a theory is merely a 'pragmatic' one. I have been maintaining all along that such an approach to theory is a misguided one, that ontological 'zoos' are an indispensable aspect of theory. Like Hempel, Salmon fails to distinguish between the reduction of independent phenomena and prediction bases in the workings of a theory. As a result, he fails to realise how the latter, *H**, *depends on the former*, *H*, in the manner described in the last section. Contrary to what Salmon contends, how we fix our prior probabilities depends on the ontological aspects of a theory.

In order to bring this out, let us go back to the fictional case of coloured, shaped objects discussed in the last chapter.[20] We can embellish our example with a few restrictions on the kinds of measurements we can make. Again, we are, at first, just measuring colour and shape aspects of things. Keeping in strict analogy to many actual scientific situations, let us assume that these objects cannot be directly observed. Furthermore, for reasons of simplification, let us stipulate that only *one* measurement be made at a time. (This is equivalent to drawing a single object from an urn and replacing it.) We can imagine our measuring instruments mounted on boxes, and each box with its own set of 'buttons' while the boxes are prepared with the system of objects 'inside'. Although it would be nice to do so, we can not open up any of the boxes in order to see what is inside. All we can do is to press buttons and take instrumental readings. Surely Salmon and the positivists would allow for at least this. Suppose, at the beginning, our experimental arrangement consists of boxes with just two buttons on them, E_1 and E_2, which measure colour and shape respectively. In order to dramatise the fact that only one measurement can be made at a time, we add that each box blows up after a button is pressed. So if we want jointly to measure both colour and shape, we cannot do so by pressing E_1 and E_2 simultaneously. This means that if we wish to measure both colour and shape, we must construct a third device, E_3, that can do the same job as E_1 and E_2, i.e., simultaneously determine colour and shape. That this can be done, in principle, is a very strong assumption. So each box now comes with three buttons: one for colour, one for shape and one which measures colour –shape combinations. If we want to measure more aspects of nature and combinations of concurrent aspects, we must add more and more buttons to our boxes. Suppose we press just E_1 and E_2 and, as it was in the last chapter, we get 50 per cent black, 50 per cent white for E_1 and 50 per cent square, 50 per cent circle for E_2. On the basis of these results, can we predict the results for E_3 pressings since colour–shape results are composed of colour and shape? The answer, strangely enough, is 'No'. What we need in order to get the correct probability for the joint occurrence of colour and shape is a good theory, i.e., one that tells us *what* colour and shape are aspects *of*.

In order to show this, construct four hypotheses, H_1–H_4, about the 'world' inside of the boxes, two of which contain ontological assumptions which are stronger than the other two:

H_1	H_2	H_3	H_4
$^1/_2$B, $^1/_2$W	$^1/_2$□, $^1/_2$○	$^1/_2$B□, $^1/_2$W○	$^1/_6$W□, $^2/_6$W○
(half are black and half are white)	(half are squares and half are circles)	(half are black squares and half are white circles)	$^2/_6$B□, $^1/_6$B○.

H and H_4 are committed to a stronger ontology than H_1 and H_2 because they are combining different aspects of nature into a single set of objects having specific properties while H_1 and H_2 are just committed respectively to colours and shapes as isolated features. Even so, H_1–H_4 equally support the experimental results, that is, as long as only E_1 and E_2 are pressed. But what happens if we start pressing E_3, making colour and shape measurements together? According to our thesis, because H^* depends on H, our theory's predictions will depend ultimately on the ontology to which it subscribes. So our predictions for *joint* experimental results vary according to our ontological assumptions and how our theory reduces the number of independent phenomena. H_3, for example, identifies colours and shapes with aspects of black squares and white circles; H_4 embeds colours and shapes just a bit differently. As a result, H_3 predicts that half of the E_3 measurements will yield black and square while H_4 predicts 1/3 will yield both black and square. What about H_1 or H_2? If they are considered separately, no predictions can be made about colour–shape measurements. However, as it was in the last chapter, H_1 and H_2 can be combined to predict that 1/4 will be black squares. But such a joint probability distribution presupposes that shapes and colours are independent phenomena while according to H_3 and H_4 they are not. So it is no wonder that we get different predictions of joint measurements as we go from $H_1 + H_2$, H_3 and H_4.

But this throws a monkey wrench into Salmon's S–R approach to theories, for the above shows that we can not correctly calculate or assign probabilities to *joint* occurrences of events in accordance with our theory unless we first ascertain the ontology expressed by that theory, i.e., ascertain how the theory reduces independent phenomena. Without doing this, all phenomena would be equally independent and their joint probabilities would be calculated simply as the product of their individual probabilities. (Notice that this would have a diminishing effect, making their joint occurrence much more unlikely. I will come back to this in the next section.) If, on the other hand, they are not independent phenomena in that they are identified with aspects of a common ontology, their joint distributions are calculated differently.[21] These points can be summarised:

I. The probability for the joint occurrence of e_1 and e_2 is simply $\text{Pr}.e_1 \times \text{Pr}.e_2$ if and only if e_1 and e_2 are independent events.

II. If e_1 and e_2 are both aspects of the same set of objects, W_n, then the probability of their joint occurrence is simply the probability of W_n. For example, according to H_4, only 1/6 of the colour–shape measurements should yield black and circle. Of course, this entails that black and circle results are no longer independent.

I cannot overemphasise the importance of distinguishing these two ways of determining the probabilities of joint occurrences. Failing to realise the difference between ɪ and ɪɪ can lead to disastrous results when it comes to measuring the degree of safety of our technology, calculations which are becoming increasingly important in our everyday lives. An example of scientists failing to distinguish between ɪ and ɪɪ has already occurred in the United States, and with disastrous results, which could have been even worse if it were not for a quirk of fate. The event I have in mind, of course, was the Three Mile Island nuclear power plant accident which occurred in Pennsylvania on 28 March 1979. Prior to this event a hotly contested debate had been going on for years over the question of nuclear safety. One side of the dispute comprised harbingers of disaster who insisted that a nuclear core meltdown (i.e., the exposure of the core to the extent that it melts its way through the containment vessel, releasing great amounts of radiation) was very much in the offing, endangering the lives of thousands of individuals, forcing them to abandon their homes, doing irreparable harm to the environment for hundreds of square miles, etc. The advocates of nuclear industry quickly replied that although such a horrifying scenario of events was a possibility, such an occurrence would nevertheless be so unlikely, it would not even be worth worrying about, that the odds of a meltdown occurring were about equal to that of being killed by a meteorite while standing on Time's Square in New York City. Since the events at Three Mile Island, we hear no more of these 'incredibly unlikely' probability arguments. Just the same, the mistake of this type of argument should be brought out. The joint events at Three Mile Island did not occur because they were highly unlikely. If so, how were the apologists for nuclear power able to drive down the probability of a meltdown in the above argument? The answer is that they simply but erroneously assumed that the combination of events which led to a disaster were independent; hence, by ɪ, the probability of their joint occurrence goes down geometrically. In light of what we know now, such an assumption was indeed tragically mistaken. On the contrary, the ontology of the situation may render these events dependent, requiring that our calculations for their joint occurrence be based on ɪɪ, leading to a much greater chance of a disaster in comparison to those calculations based on ɪ. (In fact, one analyst revealed that luck played a major role in avoiding a meltdown. Contrary to what we were being told by the 'experts', it could have gone either way.[22]) It is true that error was a major cause of the failure but what is human error, after all, but a human being *causing* the reduction of a number of independent phenomena, making their occurrence together more likely? In relation to this, the statistical computer simulation models did not take into account the possibility of a 1000 cubic foot hydrogen bubble, the presence of which turned supposedly independent phenomena into dependent ones. The moral of the story is that none of these statistical arguments are worth anything unless they have a relatively complete ontological catalogue, which was absent in their calculations. Logical positivism argues that ontology should not enter into any scientific reasonings. So, not only is a positivistic outlook mistaken, it may even be dangerous to our health!

The same point holds equally well for probabilities concerning the disjunction of events. Suppose we wanted to know the probability that a measurement will yield either black *or* square. Again, we can not answer such a question until we first determine the ontology of our theory. This becomes obvious once we inspect the formula for calculating such probabilities: $\text{Pr.}(e_1 \cup e_2) = \text{Pr.} e_1 + \text{Pr.} e_2 - \text{Pr.}(e_1 \cap e_2)$. Since $\text{Pr.}(e_1 \cap e_2)$ is a function of the theory's ontology, so is $\text{Pr.}(e_1 \cup e_2)$. There are three cases to consider here. (Remember, we are assuming that only one measurement can be made for reasons of simplicity.)

III. If e_1 and e_2 are occurrences of mutually exclusive properties of the same W_n objects, then $e_1 \cap e_2 = 0$ and $\text{Pr.}(e_1 \cup e_2) = \text{Pr.} e_1 + \text{Pr.} e_2$. So, according to H_3, if the third button is pressed, we will get either black or white, square or circle, black or circle, white or square with a certainty of one; not so, H_4.

IV. If e_1 and e_2 are occurrences of properties which are distributed among W_n objects, then $1 > \text{Pr.}(e_1 \cap e_2) > 0$ and $\text{Pr.}(e_1 \cup e_2) < 1$. Observe that the more objects that share e_1 and e_2, the smaller $\text{Pr.}(e_1 \cup e_2)$.

V. If e_1 and e_2 are occurrences of properties that have the same extensionality over W_n objects, then $\text{Pr.}(e_1 \cup e_2) = \text{Pr.}(e_1) = \text{Pr.}(e_2) = \text{Pr.}(e_1 \cap e_2)$. So, by H_3, the probability of getting black or square, black and square equals 1/2.

Just as important, conditional probabilities are determined by the way a theory's ontology reduces the number of independent phenomena, which leads us to three more principles of calculating probabilities of events. The conditional probability for event e_2 given that e_1 has occurred, Pr. (e_2/e_1) = Pr. $(e_1 \cap e_2)$/Pr. (e_1). We know by now that the probability of the joint occurrence of e_1 and e_2 is a function of how e_1 and e_2 are identified with aspects of a common ontology; so the conditional probability of e_2 given e_1 is likewise affected. On this basis we have:

VI.   If e_1 and e_2 are occurrences of mutually exclusive properties of the same W_n objects, then $e_1 \cap e_2 = 0$ and the Pr. $(e_2/e_1) = 0$. So, according to H_3, the odds of getting a black object given that it is circular are zero.

VII. If e_1 and e_2 are occurrences of properties which are distributed among W_n objects, then $1 >$ Pr.$(e_1 \cap e_2) > 0$ and Pr. (e_2/e_1) varies accordingly. For example, the conditional probability of getting a white reading given that we have already gotten square is $1/6/1/2 = 1/3$ according to H_4 while the probability of black given square is $2/6/1/2 = 2/3$. So, the more objects that share e_1 and e_2, the greater Pr. (e_2/e_1).

VIII. If e_1 and e_2 are occurrences of properties that have the same extensionality over W_n objects, then Pr.$(e_1 \cap e_2)$ = Pr. (e_1) = Pr.(e_2), which means that the Pr.(e_2/e_1) = 1. So, according to H_3, the probability of getting black given square is equal to 1. Notice how, in this case, the conditional probability equals 1 no matter how low the initial probability, Pr. (e_1) may be. In fact, a case of note is that even if Pr. $(e_1) = 0$, Pr. (e_2/e_1) still equals 1! So, by H_3, even if it were impossible for some reason to draw a square, the odds of getting black given that we were to draw a square would be equal to 1.

How does the above elementary exercise in calculating joint, disjunctive and conditional probabilities affect S–R? In the first place, Salmon fails to realise that there are at least two sets of probabilities to consider: one for our ontological zoo while the other pertains to phenomena considered independently. Principles I–VIII reveal how the former determines the latter. In order for us to determine the joint, disjunctive and conditional probabilities of events on the basis of our theories we must have a mapping, H^*, between the above-mentioned sets of probabilities. But we cannot get such a mapping unless it is one induced from H, embedding independent phenomena into a common ontology; otherwise, our probabilities *do not come from our theories* but are determined *ad hoc* by observing the frequencies of events.

We may be able to predict joint occurrences on the basis of these observed patterns but I hardly think that we have, in so doing, supplied an account of them. Redescribing them as aspects of a common ontology *does*. So contrary to S–R, explanation is more than prediction or the assignment of probabilities to events.

Since developing S–R, Salmon has come to admit that probability assignments are not sufficient to explain events. His new position is an adaptation of Reichenbach's view that theories explain by supplying one with a *common cause* underlying events. Put in the terminology developed in this chapter, events are thought of as independent phenomena and the joint occurrence of any two is to be explained by citing an underlying common cause. The two occur together because they are both caused by a third. In a presidential address entitled, 'Why Ask, "Why?" ', Salmon says that we 'have a bona fide explanation of an event if we have a complete set of statistically relevant factors, the pertinent probability values, *and* causal explanations of the relevant relations'.[23] While I think that the Reichenbach–Salmon position on explanation is definitely a step in the right direction, I also feel that it does not go far enough, for causal accounts are just a special case of reducing the number of independent phenomena. It is true that one way to explain is to supply a cause but it is not the *only* way. In the Flatland case, it is highly intuitive that the account of the occurrence of the circles in Flatland was not a causal one. This brings out an important contrast between $H–H^*$ and Salmon's new position. If Reichenbach and Salmon are right, then the relationship between independent phenomena and their underlying ontology is a causal one. On this, I disagree. According to $H–H^*$, the relationship is not causal but one of identity. The conic sections of a sphere and a plane do not cause the circles in Flatland; they *are* the circles. If two things are the same they can not cause each other.

This does not mean that $H–H^*$ does not allow for causal explanations. It most certainly does, only the causal relations do not occur between independent phenomena and the underlying ontology. Using the model of causation outlined in Chapter 3, recall that causation takes place by means of a transference of quantities between objects. Independent phenomena, here, would be the isolated movements of bodies. These movements would then be identified with the stages of momentum and energy transference. For example, in the case of body a causing b to move, a and b are identified with momentum carriers, where the former gives up its momentum to the latter. The movements of a and b are independent phenomena in that one could occur without the other. Once they are identified with momentum carriers, the conservation of momentum applies to a and b, so a has to transfer its momentum and energy to b, i.e., cause it to move. But this does not mean that the movements of a and b are caused by some other momentum carrier; they *are* the possessors of momentum! Not everything that goes on among the objects of an underlying common ontology can be said to be a transference of quantities. Lots of other things happen. Contrary to a very popularly held belief, not all explanations are causal in the mechanical sense of that term. In fact, I would venture to say that many of them are not.

Even if independent phenomena were causally related to aspects of a common ontology, there are still difficulties which plague Salmon's adaptation of Reichenbach's view, for it essentially combines S–R with a common cause, and such a joining of the two leads to internal tensions. For example, why make S–R part of the essence of explanation if the statistically relevant relations must be given a causal account? In so revising S–R, he has simply tacked on a causal explanation, which is the real core of his analysis. What he should really do is drop S–R and the structural identity hypothesis all together. His own paradigm case of a causal explanation appears to bear this out. Suppose we wish to explain, using Salmon's example, why GI Joe died of leukemia.[24] According to Salmon the explanation is in terms of S–R plus a causal account, namely, GI Joe was in the vicinity of an atom bomb blast and the subsequent gamma radiation caused his leukemia *and* the odds of his getting leukemia under these conditions were .01. But what does such a probability add to the explanation that the causal account has not already covered? (This is exactly parallel to the bullet-randomised field case.) Radiation caused a change in his cell structure which was later manifested as leukemia. What does saying that such an event was highly unlikely add? This is another way of saying that causal relevance and statistical relevance do not go hand in hand, as they answer entirely different questions. So we are back to his equivocation on the explanandum again. It is one thing to ask why (or how) GI Joe got leukemia but quite another to ask why a very small percentage of soldiers who were present at the atom bomb blast got the disease. A .01 probability for the occurrence of that particular causal process is a relevant response only to the latter question if it is appropriate at all. (It does not seem to explain *why* – it just repeats the question.)

In conclusion, while S–R proponents believe that a theory's ability to explain depends on how it predicts or assigns probabilities to events, I have argued that it is the other way around, that how a theory predicts is a function of how it explains.[25] This provides the key to freeing ourselves from the positivist approach to science and places us in a position to appreciate fully the essential role theoretical entities have in science. We are ready, then, to complete the argument for an ontological approach to theories which was outlined at the beginning of Chapter 6.

5. WHY LOGICAL POSITIVISM IS A POOR BET

Can metaphysics be completely purged from scientific theory? If what I have been arguing for all along is correct, not only is the answer to this question 'No!' but metaphysics plays a predominant role in theory in that theories explain and predict as a function of the way they refer to entities such as photons, electrons, forces, fields, etc., some of which are impossible to observe. While I realise that this would be a bitter pill for many philosophers and scientists alike to swallow, the arguments presented here are meant to show that this is the only route to go if we are to make sense of how a working science is able rapidly to confirm its hypotheses. Below, I will utilise the analysis of confirmation in Chapter 6 and the $H–H^*$ model of explanation to argue that realist theories are more open to confirmation than those which fit positivist models *because of their reference to theoretical entities*.

Logical positivism seeks completely to expel metaphysics from theory. But, as Wittgenstein realised when he wrote the *Tractatus*, to maintain that the world consists of nothing but the facts, that each atomic fact is independent of every other fact, and that there is no such thing as a causal nexus, *amounts to committing oneself to some kind of an ontology* albeit a very 'weak' one consisting of independently occurring phenomena which are covered by universal conditionals (or laws). This does not mean, then, that logical positivism is above scientific scrutiny. On the contrary, whether the logical positivist is willing to admit it or not, his stance is not just a meta-theoretical one; it contains a definite view on the nature of things: namely, the maximal ontology permissible in a scientific theory is one where all phenomena are events which occur in accordance with laws, and that they are not identifiable with anything theoretical underlying them such as atoms, fields, quarks or whatever. Their ontology stops at the observable, event level and nothing beyond is to be of genuine scientific interest. It is true, according to the positivist, that scientists may insist upon using theoretical terms in their descriptions and explanations of phenomena but such a use is a *conventional* one, reflecting a *choice* of words rather than a successful reference of theoretical terms to their corresponding objects. Recall the discussion of conventionalism in Chapter 2. The conventionalist claims that different theories with the same subject matter are not really rival theories; they are simply different ways of 'locating' the facts within a conceptual scheme. Which scheme is to be used to pick out the facts is a matter of 'taste' or pragmatic considerations alone.

Even if logical positivism is committed to this bare, minimal ontology, it is an ontology nevertheless. What remains to be done, then, is to see how such a view of the nature of things goes about explaining and fixing probabilities for combinations and permutations of independent phenomena. Certainly, the positivist's view of nature, then, is not immune from comparison with the realist's hypothesis that there are entities underlying independent phenomena. The question now becomes: *which hypothesis is more open to confirmation*, that is, *which hypothesis about the nature of things does a better job of accounting for combinations and permutations of independent phenomena?* It has been my contention all along that the realist hypothesis fares much better. As a result, according to the analysis of confirmation presented in the last chapter, scientific realism (= the class of theories which refer to entities underlying observable, independent phenomena) is more confirmed than logical positivism (= the class of theories where nothing but observables are objects of reference). But the logical positivist has always insisted that the best theory is the one that is better confirmed over its rivals. Thus, by the logical positivist's very own standards, theories which predict and explain in virtue of their reference to entities underlying (or beyond) observables are better than theories that refer to observables alone.

Let us see how all of this works, making use of the analysis of confirmation in Chapter 6 and principles I–VIII in the last section. (The reader is urged to review Chapter 6.) Now, the logical positivist gives us this much: each set of independent phenomena consists of nothing but observables, observables which behave in accordance with laws. Returning to the cough–fever case, it was argued that these independent phenomena were reduced by identifying them with aspects of an underlying and, supposing for the sake of argument, unobservable microscopic, biochemical process. In doing so, we realised that they are not independent phenomena after all, i.e., in accordance with the above two senses of 'independent'. Once we realise how it is that their existence together is not coincidental, they are understood. Right away, the positivist will reply that their appearance in the hapless victim can be equally accounted for by replacing the theoretical microprocess with a law which states that under such and such medical circumstances, a slight fever will be accompanied by a chronic cough. True enough but notice that the above microscopic process has *other aspects*, that it is many-faceted by its very nature. So, it will have things to say, for example, about the frequency of coughing in relation to the fever of the victim, the amount of perspiration in relation to the air capacity of the lungs, the degree of pupil dilation as a function of pulse rate, and so on, *ad infinitum*. (Of course, I am making all of this up for purposes of illustration. I hope medically knowledgeable readers will forgive my naïvety.) However, according to logical positivism, there is nothing for the cough and fever of the victim to be aspects *of*. So, even if there is a known law linking up the existence of the fever and the cough, what could possibly motivate formulating all of these other relationships in terms of the coexistence of a variety of symptoms, as well as functional relationships among their values? While the biochemistry of the disease may tell us to expect an exact correlation between pupil dilation and pulse rate there is nothing in the lawlike relationship between the cough and fever that would tell us *that*. Whereas reference of the cough and fever to a common, underlying ontology of a disease will generate these new relationships, the positivist will have to wait for their (unlikely) discovery by observation, which is done by trial and error. In other words, while a common ontology supplies one with good reason to expect a variety of phenomena to occur in specific combinations, the *only* reason a positivistically based hypothesis can supply is that these combinations have been *separately observed*.

The above is mainly a methodological point which leaves the positivist with this out. He can always claim that theories are just convenient myths which can be used to generate all of the new lawlike relationships among observables. Once we get them, he will insist, we can always discard the common ontology that generates them. This is a very familiar move, one which is similar to the claim that models were dispensable. While I agree that, in some sense, a long string of lawlike relationships can be substituted for theoretical entities, a very high price must be paid by the positivist for doing so, a price that even he cannot afford to pay. Notice how the above ploy resembles the attempt to save the Flat Earth hypothesis which was discussed in section 4 of the last chapter. Recall how that hypothesis was saved by formulating a series of (*ad hoc*) correlations that were intended to do the same job of handling data as the Round Earth hypothesis. Even though each correlation handles the data it was restricted to, when this long list of correlations was tested against the Round Earth hypothesis across categories, the Flat Earth hypothesis lost in confirmation at a geometrical rate. I maintain that logical positivism is open to the same kind of attack that was levelled against the Flat Earth hypothesis.

Once he denies the existence of a common, underlying ontology, the positivist denies himself the use of the laws of that ontology in order to generate probabilities for permutations and combinations of independent phenomena. For example, he cannot use laws which cover the unobservable disease to generate all the laws covering observable symptoms if he denies the existence of the disease in the first place. Instead, these probabilities must result from a long list of *independently maintained* laws among observables, one for each observed relationship among permutations and combinations of independent phenomena. So, even if he has a set of lawlike relationships between the cough and fever, perspiration and air capacity of the lungs, pupil dilation and pulse rate, there are still more combinations to consider, say, between coughing and pulse rates, degree of fever and pupil dilation, etc. Whereas *all* of these combinations follow quite naturally from the biochemical features of the unobservable disease, they must each be tacked on to the original list of laws covering observables. In other words, while it is built into the unobservable aspects of the disease to produce these new relationships without end, the positivist must add these new relationships to the old list if his version is to generate this new data at the same rate. Recall that an important spin-off of identifying independent phenomena with an underlying ontology is that lawlike relations going *across* independent phenomena are generated. The functional dependence between the temperature of a gas and the solidity of ice, the rate of a body's fall and the area swept out by a planet were such cases in point; and these are precisely the kind of cases, I maintain, which cannot be anticipated by positivistic theories.

There is good reason why the positivist finds himself in this quandary. Because the realist *identifies* independent phenomena with aspects of a common ontology, i.e., uses the identity relationship in his account of them, he is able to utilise Leibniz's indiscernibility of identicals to transfer the laws which cover various aspects of a common ontology to independent phenomena. To repeat what was already said, any laws covering different aspects of a common ontology must also cover independent phenomena (and vice versa) if the latter are identified with the former. Now the positivist does not have this principle available in his account of independent phenomena because they have nothing 'below' with which to be identified. For this reason, common causes cannot account for relationships that go across categories either. If independent phenomena are causally related to their underlying ontology then the indiscernibility of identicals cannot be used to transfer lawlike relationships from the latter to the former. Only the identity relationship can generate these cross-category laws. So, even if some third thing caused the cough and fever in the bronchitis victim, we are still at a loss as to why the frequency of coughing and the degree of fever are so related.

We can now compare the realist and positivist ontological points of view with respect to confirmation. What kind of data is to serve as a test between them? It will concern the frequency of occurrence of certain combinations of different types of independent phenomena, as well as relationships that hold between their values. It would be like the variations on the black–square, white–circle cross-categories case which we went over in the last section. Principles I–VIII tell us in what way the probabilities of different combinations of independent phenomena are a function of how they are reduced by the objects of an underlying ontology. One result that comes out of this is that *to the extent that an underlying ontology permeates independent phenomena, the probability of certain cross-category combinations of occurrences should be quite high relative to other combinations.* So, if the odds of getting bronchitis (characterised in biochemical terms) are 1/2, the *odds of the total of every one of the above functional relationships among the symptoms are also* 1/2. Once the individual gets the disease, the odds of all of these symptoms occurring, along with their interrelations, are extremely high, approaching 1.

Suppose, then, according to our common ontology and its laws $(L_1, L_2, L_3, \ldots, L_n)$, the probability of occurrence of an indefinite number of cross-category, functionally related combinations of independent phenomena is .9. Letting E_n stand for the totality of independent phenomena across categories, the maximum likelihood table for this case would be:

$$\Pr(E_n/H)\frac{\theta_1 = \text{common ontology} + L_1, L_2, L_3, \ldots, L_n}{.9}$$

How does logical positivism handle the very same data, E_n? I said earlier that a common ontology and its laws must be replaced by an indefinitely long string of laws over observables. For the sake of argument, let us say that each one of these laws has, in fact, been observed to be true to a degree of .9. There are two kinds of laws to deal with in this case. One set of laws covers independent phenomena within a category (L^{IP}) while the other covers relationships *across* categories (L^{AC}) or across independent phenomena. Letting 'IP' stand for observable, independent phenomena, this, then, is the table for logical positivism:

$$\Pr(E_n/H)\frac{\theta_2 = (IP + L_1^{IP}, L_2^{IP}, \ldots, L_n^{IP}) + L_1^{AC} + L_2^{AC} + \ldots L_n^{AC}}{.9 \qquad\qquad X.9 \quad X.9 \ X. \ .X.9}$$

As it was in the case of the Flat Earth hypothesis, because the L_n^{AC} are independently maintained, the probability of E_n given θ_2 is driven down at a geometrical rate. Whereas a common ontology tells us to expect many, many functional relationships among independent phenomena over diverse categories, the universe of the logical positivist has us expect a uniform but very *minimal number* of cross-category functional relationships. This, again, follows from I–VIII. Recall how in the colour–shape examples, the probability of colour–shape combinations decreased when they ceased to be considered features of underlying objects.

Conversely, if the universe had a scant number of cross-category correlations of events and their values, the positivist approach would have the better of it because that is precisely what it predicts and the opposite of what falls out of a realist ontology. So, all that has to be done is to look at actual scientific progress in order to determine whether lawlike relationships of the cross-category kind have occurred or not. The answer is that they exist in abundance! In physics, for example, as we progress from one theory to its successor, more and more of these relationships come out. This brings us back to the discussion of the rapid confirmation of the general theory of relativity in the last chapter. Recall how so many relationships across diverse phenomena were generated from GTR. Temporal, geometrical, gravitational and dynamical phenomena were unified as never before. Even biological processes such as the rate an organism ages relative to others as a function of the velocity of its inertial frame of reference are now captured by the net of GTR. One wonders how such an 'outlandish' relationship could have been predicted by working with biological and physical observables alone.

It is not convention, then, which motivates postulating theoretical entities. Without them, we could not have anything on which to hang independent phenomena, and we could not make sense of the rapid accumulation of cross-category relationships. It is not pragmatic dictates, then, but the fact that realist theories do a better job of accounting for this data and are, hence, more confirmed. Their ontology is more suited for these cross-category results. Thus, the positivist ontology is no better off in terms of confirmability than the Flat Earth hypothesis.

In answer to the question 'Do we have any reason to believe that our theories are really telling us about nature, that they are actually referring to a reality beyond the world of common-sense, everyday observables?', I say we do have good reason, namely, the only way we can possibly make sense of the rapid accumulation of the above type of cross-category relationships is that our theories are successfully capturing a reality underlying observables. This answer is well in keeping with the realist tradition, which insists that the success and progress of science is the best evidence of all for the existence of theoretical entities. Putnam expressed this point well: 'The positive argument for realism is that it is the only philosophy that doesn't make the success of science a miracle.'[26] This kind of argumentation can be traced to Boyd's work, where he argues that realism, as opposed to positivism or conventionalism, is the only way to account for the systematic experimental success of a certain class of theories over its rivals. Unlike the realist, the conventionalist is at a loss when it comes to explaining how it is that theories which presuppose other, auxiliary hypotheses about causal relations among theoretical objects are constantly winning out over their rivals that go against these auxiliary hypotheses.[27]

Not that the debate between the realist and the positivist has ended once and for all. Certainly, this writer has no expectations that the above argument will put an end to the controversy. Even at this writing anti-realists such as van Fraasen and Laudan have recently attacked the above realist line of argumentation.[28] Unfortunately, a detailed discussion of their work is beyond the scope of this text. The more advanced student is advised to study them in order to compare and contrast the strengths and weaknesses of each position on the question whether science can operate on a purely empirical level, free of any metaphysical assumptions whatsoever. The argument which has been developed in Chapter 6 and this chapter is designed to show that not only can metaphysics in the form of ontological commitments not be purged from theory, the ontological commitments of a theory, in fact, comprise its very essence.

One of the most important types of scientific explanation occurs when one scientific system is reduced to another, more fundamental one, when the results of the former are understood in terms of the latter. It should not be too surprising to the reader to learn that this type of explanation is a special case of explaining by reducing the number of independent phenomena. Let us now turn to reduction in the sciences.

NOTES AND REFERENCES

1. A more recent explication of this approach can be found in Rom Harré, *The Philosophies of Science* (London: Oxford University Press, 1972) pp. 100–37; 168–83.

2. In *Statistical Explanation and Statistical Relevance* by W. Salmon, R. Jeffrey and J. Greeno (University of Pittsburg Press, 1971).
3. Ibid., p. 11.
4. Ibid., p. 77.
5. Ibid., p. 78.
6. Ibid., p. 79.
7. Ibid., pp. 57–8.
8. Ibid.
9. Salmon, *The Foundations of Scientific Inference*, pp. 124–32.
10. Many of the ideas in these sections came from research I did in collaboration with R. Weingard and D. Edwards.
11. William Kneale, *Probability and Induction* (London: Oxford University Press, 1949) pp. 87–97.
12. Michael Friedman, 'Explanation and Scientific Understanding', *The Journal of Philosophy*, vol. LXXI, no. 1 (1974) pp. 5–14. Strictly speaking, this is not true, as these laws can not be derived from Newton's laws alone. This point will be covered in the next chapter.
13. P. Kitcher, 'Explanations, Conjunctions, and Unification', *The Journal of Philosophy*, vol. LXIII, no. 1 (1976) pp. 207–13.
14. J. L. Austin, *Sense and Sensibilia* (London: Oxford University Press, 1962) pp. 68–70.
15. Formulating precise criteria for the identity of events is a demanding task, indeed, with many attempts having occurred in the literature. Listing just a few of them: (1) M. Bradie, 'Adequacy Conditions and Event Identity', *Synthese*, 49 (1981) pp. 337–74; (2) M. Brand, 'Identity Conditions for Events', *American Philosophical Quarterly*, 14 (1977) pp. 329–37; (3) D. Davidson, 'The Individuation of Events' in N. Rescher (ed), *Essays in Honor of Carl A. Hempel* (Dordrecht: Reidel, 1970) pp. 216–34; and (4) J. Kim, 'Events and Their Descriptions: Some Considerations', in Rescher, pp. 198–215.
16. E. Abbott, *Flatland* (New York: Dover Press, 1953) pp. 76–83.
17. For an excellent exposition on this, see N. Calder's *The Key to the Universe* (New York: Viking Press, 1977). This example was taken from there, as I believe it perfectly fits explaining by reducing the number of independent phenomena.
18. Ibid., p. 73.
19. D. Freedman and P. van Nieuwenbuizer, 'Supergravity and the Unification of the Laws of Physics', *Scientific American*, vol. 238, no. 2 (1978) p. 139.
20. I owe this example to D. Edwards.
21. Ironically, it was H. Reichenbach, a positivist, who was one of the first to notice that ontological features (in particular, common causes underlying phenomena) can greatly affect the assignment of joint probabilities to events. See his *The Direction of Time* (Berkeley: University of California Press, 1956) pp. 157–67. I say 'ironically' because he did not realise how such a connection could be instrumental to what I take to be a devastating realist attack on positivism, which will be presented below.
22. E. Marshall, 'A Preliminary Report on Three Mile Island', *Science*, vol. 204 (1979) pp. 280–1.

208   *A Realist Philosophy of Science*

23. Salmon, *American Philosophical Association: Proceedings and Addresses*, vol. 51 (1978) p. 699.
24. Ibid., pp. 688–9.
25. Another version of the connection between how a theory explains and the way it predicts and gathers evidence can be found in G. Harman's view that evidence is inference to the best explanation. (See Chapter 8 of his *Thought* (Princeton University Press, 1973).) However, as P. Achinstein points out in *Law and Explanation* (London: Oxford University Press, 1971) pp. 119–22, while his story depends on the notion of explanatory power, he never supplies us with an account of what it is. Likewise, he fails to distinguish between explanatory power and the probabilities for events that theories generate. *H* and *H** clearly make this distinction while showing exactly how the nature of a theory's evidence depends on the way it explains.
26. H. Putnam, *Philosophical Papers* (London: Cambridge University Press, 1975) p. 73.
27. R. Boyd, 'Realism, Underdeterminism, and a Causal Theory of Evidence', *Noûs*, vol. VII, no. 1 (1973) pp. 1–12.
28. Respectively in *The Scientific Image* (New York: Oxford University Press, 1980) and 'A Confutation of Convergent Realism', *Philosophy of Science*, vol. 48, no. 1 (1981). It is interesting to note that Laudan agrees with the realist on the matter of pitting realist theories against positivistic ones but he contends that realism will lose the bet in that it 'is neither supported by, nor has it made sense of, much of the available historical evidence' (p. 20).

8   The reduction of theories

The kinetic theory of matter was by far one of the most exciting achievements in the history of science. Not only did it succeed in explaining a vast array of hitherto mysterious macroscopic phenomena by reducing gases, liquids and solids to swarms of microscopic atoms whose behaviour is governed by Newton's laws and the laws of statistics, it also enabled the incorporation of independent or isolated scientific systems into a single, more comprehensible and powerful framework. So the reduction of theories augments our understanding of nature in a very special way, and in such a way as to bring about a single coherent picture of nature, something that the scientific community has always sought to achieve from the very beginning, ever since Thales boldly asserted that everything was (a form of) water. This venerable way of accounting for natural phenomena has been the subject of philosophical treatment by philosophers of science for ages. Although, at first, it appeared that analysing the logic of explanation by reduction and the relationship between the reduced theories and the deeper system to which they are reduced would be a fairly straightforward matter of applying traditional techniques of logic and philosophy, it turned out, on the contrary, that the logic of reduction has been highly elusive, and that the relationship between the reduced theories and the reducing theory is a far more complex one than was first envisaged. As a result, controversy over these issues has persisted in the literature up to the present.

1.   THE TRADITIONAL D–N TREATMENT OF THEORY REDUCTION

In Chapter 3, I mentioned that Hempel held that the reduction of one theory in terms of a deeper one comes about by means of deduction: the laws and results of the reduced theory are to be logically derived from the laws and fundamental assumptions of the reducing theory.[1] And, of course, according to the tenets of D–N, such a derivation automatically becomes an explanation of the laws and results of the reduced science.

As tidy as the D–N picture of reduction may have appeared to those who maintained that reduction was achieved by means of deduction, it became apparent that modifications were in order; for even though the logical derivation of the reduced theory from the reducing theory appears to be a reasonable programme on paper, a difficult problem arises when it comes to such a derivation in practice.[2] The source of this problem lies in the fact that terms appear in the laws of the reduced science which can not be found in the lexicon of the reducing theory, rendering the above logical derivation invalid. For example, according to the standard account of reduction, if we are to give a microscopic explanation of the macroscopic data embodied in the empirical gas law, $PV = N k T$ (the Boyle–Charles law, where 'P' stands for pressure, 'V' for volume, 'T' for temperature, 'N' for the number of particles, and 'k' is a constant) in terms of the kinetic theory of matter, we must deduce the empirical gas law from the laws of kinetic theory. Well, if we try to do this, the above problem becomes obvious. Let us start with the laws of statistical mechanics and see how far we get. According to kinetic theory, a real gas is conceived of, as a first approximation, as an ideal gas. An ideal gas is composed of (again, our ontological zoo) nothing but infinitesimally small, perfectly elastic particles whose only mode of interaction is by collision. Thus an ideal gas has only mechanical properties. On the statistical assumptions of kinetic theory, the pressure an ideal gas exerts on the walls of its container is expressed by the equations

$$(1) \qquad P = (2/3)\, n\, <mv^2/2>$$

('n' stands for the number of particles per unit volume while '$<mv^2/2>$' for the mean kinetic energy of the particles). Multiplying sides of (1) by volume, we have

$$(2) \qquad PV = (2/3)Vn <mv^2/2> = (2/3)\, N<mv^2/2>$$

While (2) is beginning to look like the empirical gas law, it will never get there, because 'temperature' is not a term to be found in statistical mechanics. We must add another premiss, (3), to (1) and (2) in order to close the logical gap

(3)   ???

How is (3) to be characterised?

There have been two approaches to the above problem. The first is semantical or definitional. If we define 'temperature' in terms of the mean kinetic energy of a gas, i.e., 'T' *means* $2/3$ $k < mv^2/2 >$, we then have

(3) $T = 2/3k <mv^2/2>$

and $P\ V = N\ k\ T$ follows immediately. Reduction, here, is reduction in meaning. Among some of the difficulties with this approach is that reduction comes about much too easily. Can we not always take two sciences and derive the laws of one from the other simply so by making up definitions? The problem is that the reduction of a theory is a significant scientific achievement not a semantical one. Deriving by definition introduces an arbitrariness and artificiality into reduction which does not do justice to actual scientific accomplishments.

The second technique of deduction is known as the 'bridge laws' approach.[3] Instead of interpreting (3) as a definition, it really stands for a lawlike relationship between temperature and mean kinetic energy. (3), then, is considered to be an empirical law that bridges the two sciences; and once lawlike propositions like (3) are added to the laws and assumptions of the reducing theory, derivation of the reduced theory immediately ensues. Thus, the advent of bridge laws enables the application of D–N to reductionistic explanations.

There are problems with the bridge laws story of reduction as well. There is reason to suspect that bridge laws are not so empirical as it was intended by their authors. So long as bridge laws like (3) are conceived of as statements which enable the derivation of macroscopic laws from a microscopic theory, their adequacy is actually determined on logical rather than experimental grounds. The reason why

$T = 2/3\ k\ <mv^2/2>^2$

is an inadequate bridge law is not that it is empirically false but that it does not lead to the Boyle–Charles law. Even worse, recall that we experimentally confirm our hypotheses by combining them with auxiliary hypotheses in order to generate predictions. But combining (3) with (1) and (2) only leads to an already well-confirmed Boyle–Charles law. How, then, can bridge laws be falsified if they always lead to true consequences? Something is definitely wrong here in that, like the definitional approach, bridge-laws lead to an arbitrariness and artificiality which belittle a major scientific feat.[4]

But most important, bridge laws fail to reduce theories. After all, bridge laws simply connect or relate properties in different sciences, correlations being the strongest relation between them. However, the notion that the laws of one science are derivable from the laws of another science by means of bridge laws is perfectly compatible with anti-reductionist views on theories. One well-known, extreme case of an anti-reductionist stance is Cartesian dualism, which insists that the mental can not be reduced to the physical in that mental and physical states are separate and distinct; yet, why can't a dualist still allow for there being correlations (= bridge laws) between mental and physical states in such a way that the laws of one science can be derived from the other? They are still separate sciences just the same. Those who believe in emergence, i.e., maintain that some properties of living systems can not be reduced to the micro-properties of complex inanimate groups, can take a similar stance: this time, the laws bridge properties which are on different levels. So bridge laws do not appear to be enough to warrant reduction.

What I will show below is that the reduction of scientific theories can be achieved without having to postulate either definitions or bridge laws. On the contrary, propositions like (3) are the *result* of reduction as opposed to being a fundamental assumption of it. In opposition to the traditional approach, I again oppose the view that reduction is characterised as an inference from the reducing to the reduced theory. Keeping in accordance with the approach to explanation presented in the last chapter, I feel that the most promising approach to revealing the nature of theory reduction will be that of my ontological approach to theories.

2. ONTOLOGICAL REDUCTION

I have previously argued that the explanatory power of a theory lies in the ontology it presents to us, and since different theories express different ontologies, it is my contention that reduction of theories takes place on the ontological level; as a result of reduction, the kinds of things and properties of things there are in the universe end up being fewer. For example, atomism is a reductionist theory which says that a variety of objects such as tables and chairs, rainbows, people, etc. are really one kind of entity, namely, configurations of atoms. Likewise, the properties of tables and chairs, rainbows, mental states, etc., are really one set of properties which atoms and configurations of atoms possess. So theory reduction has us view the world parsimoniously by reducing the number of things and properties.

How is this done? How can we even *say*, 'Before there were many but now there are few'? The answer to this, once more, can be found in the identity relationship, '— = —'. If we use lower case letters as names to represent entities, we can say

'$a = b$' means that a and b refer to or denote one and the same thing.

For example, 'The Morning Star = The Evening Star' means that the above two names refer to one and the same heavenly body. 'John Smith' and 'Slim' may be different names of the same person while 'Jimmy Carter' and 'The 39th President of the United States' are a name and a definite description that refer to the same individual. This also holds for the properties of things. Different predicate expressions may actually refer to one and the same property. Using upper case letters for predicates,

'$P = Q$' means that the predicates P and Q stand for (or refer to) one and the same property.

So, for example, 'Red' and 'The colour of blood' are predicate expressions that denote the same property or colour. A more important example concerns the relationship between the temperature of a gas and the mean kinetic energy of the gas molecules according to kinetic theory. Although the meaning of 'temperature' is independent of the meaning of 'mean kinetic energy', one striking result of reducing classical thermodynamics to statistical mechanics is that 'temperature' and 'mean kinetic energy' are predicate expressions or variables that actually refer to one and the same property or state of an ideal gas.

The identity relation tells us that what may appear to be different because of the use of different names and predicates is really the same. Why must there be a one-to-one correspondence between an object and a name? '=' checks us from proliferating entities and their properties by a 'semantical subterfuge' of adding names and predicate expressions to our vocabulary. We may wish to know, for example, how many guests will be present at a party by inspecting the invitation list. If there are twenty names, does this necessarily mean that twenty people were invited? Suppose we learn that two of the names on the list actually refer to the same person. We are thus down to nineteen guests. This then, is how the identity relation expresses ontological reduction. In the beginning, several independent theories originate in such a way that it appears they are committed to a wide range of ontologies. (For example, we can imagine a spectrum of theories, ranging from microphysics to economics and psychology.) A successful reduction, however, involves showing how these theories are actually committed to the same ontology. They are simply describing the same realm of entities and properties in different ways just as two names may end up referring to the same individual in different ways. Similarly, the programme of kinetic theory is committed to the idea that the only properties of an ideal gas are mechanical properties. But we know that gases have thermodynamical properties as well, temperature being one of them. So if temperature is a property of a gas and if kinetic theory is true, then temperature will turn out to be the *same* as one of its mechanical properties, namely, the mean kinetic energy of its atoms. 'Temperature' and 'mean kinetic energy' are just different ways of denoting the same state of a gas.

Now that we have taken care of these semantical preliminaries, it can be seen why the traditional approach to the reduction of theories is so wanting. Again, the fault lies in neglecting the ontological aspects of a theory in favour of its formalistic features. There is nothing in the D–N account which expresses how a theory reduces the number of entities and properties in the world; yet, this is really what reduction is all about. The major kernel of a reducing theory is a proposition which asserts that the entities and properties which are the subject matter of the reduced science(s) are identified (the same as) with the entities and properties referred to in the reducing science. So, even though there are at least two lists of names and predicates, say, N number of properties of the reduced science and M number of properties of the reducing science, since any property of the reduced science will be identified with one of the reducing sciences, the total number of properties will *not* be equal to $N + M$ but M, thanks to the identity relationship. It should be no surprise by now why the traditional D–N approach to reduction overlooked such an important proposition of the reductionist programme. After all, the above claim that any macroentity or macroproperty is the same as a combination of microobjects and their properties is strictly ontological in nature, and it certainly can not be considered to be either a law of nature or an initial condition, as D–N most definitely requires.

These identities between the entities and properties are the true source of explanatory power in theory reduction. Contrary to Hempel, we do not explain the laws of the reduced science by deriving them from the laws and assumptions of the reducing science plus a bridge law. As I pointed out in the last chapter, an immediate consequence of establishing that two things are the same is that they share the same properties, behave in exactly the same way in all respects, have exactly the same causes and effects, etc. This was encapsulated by Leibniz's law of the *indiscernibility of identicals* and it is this principle which underlies all explanations of macroscopic phenomena in microscopic terms. How, then, should the empirical gas law be explained if not by logical derivation? I have said that the programme of kinetic theory specifies that temperature be identified with a mechanical property. (If temperature can not be so identified, then kinetic theory is false.) It so happens that the mechanical property which turns out to be the same as temperature is the mean kinetic energy of the gas atoms. The result of this identification is that we have gained new information about temperature, and it is this new information which enables us to explain why the pressure, volume and temperature of a gas are related as they are described in the Boyle–Charles law.[5] For example, since the temperature of a gas is the same as the mean kinetic energy of its gas molecules, we can explain why, at a constant volume, increasing the temperature increases the pressure: if volume is held constant, by Leibniz's law, to increase the temperature *is* to increase the average kinetic energy of the molecules; this leads to an increase in the rate of change of momentum per unit area on the container walls, i.e., an increase in pressure. If we were to apply pressure at a constant volume, this would amount to doing work on the gas which must be converted into an increase in the energy of its molecules, i.e., an increase in temperature. Likewise for the other relations mentioned in the empirical gas law. We repeat: this is not done by deriving the law but by showing how it works according to the ontological story given by kinetic theory.

3. THE LOGIC OF EXPLANATION BY REDUCTION

It can be shown how explanation by reduction is a limiting case of explaining by reducing the number of independent phenomena. Once more, (the reducing) theory explains by providing a common ontology. In the non-reduction cases of explanation, independent phenomena are identified with *different* aspects of a common ontology. For example, in our hypothetical colour–shape study in the last chapter, black and square events were identified with different facets of coloured-shape objects. In the case of reduction, however, different properties among systems of independent phenomena are identified with the *very same aspects* of a common ontology which is provided by the reducing theory. Prior to the kinetic theory, classical thermodynamics and Newtonian mechanics dealt with independent subject matters, each science with its own set of laws, properties and things. After reduction, each property in classical thermodynamics is identified with an aspect of the common ontology of the reducing theory. In classical thermodynamics, for example, the temperature of an ideal gas is identified with the mean kinetic energy *of* a collection or swarm of gas molecules. But when Newtonian mechanics is considered as a separate science, trivially, it has a phenomenon which is identified with the very same aspect of a collection of gas molecules, namely, the average kinetic energy of a collection of molecules. Thus, temperature and mean kinetic energy are identified with the very same aspects of a common ontology provided by the reducing theory.

What this means is that there are two ways that reducing

theories reduce the number of independent phenomena. On the one hand, the reducing theory provides an ontology which is *neutral* relative to independent phenomena. The number of properties and things are drastically reduced in this case, because of the transitivity of the identity relationship: if P and Q are independent phenomena which are identified with the very same aspect of a common ontology, S, then P and Q are identified with each other. On the other hand, reduction usually does not take place by providing a neutral ontology but instead by means of ontological *redundancy*:[6] the objects and properties of one of the independently held theories become the objects and properties of the reducing theory. In other words, before reduction, scientists were unaware that the ontology of one of the independently held theories could serve as an ontology common to all the others.

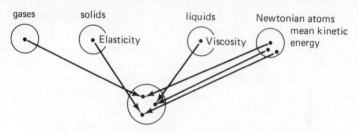

FIGURE 8.1 *Various aspects of Newtonian atoms*

Before the kinetic theory, no one dared to use the corpuscular ontology to treat various phenomena such as gases, liquids, etc. As such, they are independent systems of objects. Since the properties of gases are identified with aspects of collections of Newtonian atoms and since the properties of Newtonian atoms are trivially identified with themselves, again independent phenomena are identified with the very same aspects of a common ontology, leading to a reduction of properties and entities. And, of course, *because the identity relation is transitive, we now have identities between properties of independent phenomena.* Property reduction ensues.

Let us see how the logic of this looks in terms of H. Reduction is represented by homomorphism, H, between a representation of the reduced theory(ies) and a representation of the reducing theory. Again, the laws and relationships of each independent theory are preserved in the mapping. That independent phenomena are identified with the very same aspects of a common ontology is mirrored in H by having *the elements of the representations of the disjoint, independent phenomena systematically hit the same elements* (or, the same subsets of elements) *in the common ontology representation*. (Notice how temperature and mean kinetic energy hit the same element in the range of H in Figure 8.1.)

There are two things of note in the above mapping. Once more, the D–N approach has it wrong; reduction is *not* formally represented as a derivation of the reduced theory(ies) from the reducing theory but as a map from the former to the latter. So the traditional D–N approach to the logic of theory reduction has again confused the direction of inference. There is another, very important result that follows from the map, H, from the reduced to the reducing theory, one that will settle the bridge laws issue once and for all. Recall that bridge laws were imposed by the logician as lawlike relations between properties of independent scientific systems in order to fill a logical gap between the reduced and reducing theory. I maintained that to impose these relations in the guise of discovering empirical laws by scientific method was misleading because such a move is really alien to actual scientific practice. Besides, we no longer have to use bridge laws as derivational devices because reduction is not a derivation as the D–N proponents believed.

As was pointed out in the last chapter, a most important consequence of characterising the logic of reduction in terms of H is that lawlike relations across independent phenomena result once the latter are identified with aspects of a common ontology. Recall how this follows directly from the law of the indiscernibility of identicals: if P and Q are the same property, then P and Q will behave in exactly the same, lawlike way with respect to all other properties. So, for example, if 'temperature' and 'the mean kinetic energy of a collection of gas atoms' denote the same property of an ideal gas, then if the temperature of a gas is proportional to the product of pressure and volume, so is the mean kinetic energy of the gas's atoms. But, then, we now have several lawlike relationships – $T = 2/3k < mv^2/2 >$, for example – between parameters of classical thermodynamics and statistical mechanics in the way Newtonian mechanics generated a new lawlike relationship between the distance covered in freefall and the area swept out by a body in orbit. Likewise, if the work done on a gas is related in a lawlike way to an increase of the gas's mean kinetic energy, it is also related in the same lawlike way to an increase of the gas's temperature, a result that has had much significance in the confirmation of the kinetic theory of matter. What this means is that lawlike relationships between parameters of independent systems, our bridge laws, are not propositions that are magically pulled out of a logician's hat in order to deduce the laws and results of a reduced science from the laws and assumptions of a reducing science; rather, bridge laws are *theorems* which naturally fall out as a result of embedding the reduced theory(ies) into the reducing theory. In other words, if we have correctly construed the logic of reduction, bridge laws are not postulates of reduction, as it has been traditionally believed, but the *result* of successfully reducing the number of entities and properties originally maintained by our theories.

I have focused my discussion so far on the logic and meaning of reduction. What would count as evidence for a reducing theory? What kinds of predictions do they make? We know now that H induces H^* which takes us from the probabilities of the common ontology to the probabilities of the independent phenomena. The identities between the entities and properties of the reduced system, on the one hand, and the entities and properties of the reducing system affect H^* in this way: by principle II of the last chapter, since $P = Q$ the probability distributions for P and Q are exactly the same. This is just a statistical version of Leibniz's law of the indiscernibility of identicals. If two properties are the same then we can predict that they will behave exactly the same way in relation to all other properties. Again, if the temperature of a gas is the same as the mean kinetic energy of its molecules, then any change in temperature with respect to a given variable should have a corresponding change in the mean kinetic energy and vice versa.

I mentioned how the identity of temperature and mean kinetic energy leads to a previously unknown and highly significant relation between work and temperature. It also leads to a prediction which serves as a crucial experiment between different characterisations of heat. Is heat a fluid substance or is it a mechanical phenomenon? If temperature is identical with mean kinetic energy, since we can increase the kinetic energy of a system by performing work on it, we should thereby be able to increase the temperature of a gas by means of mechanical work alone. On this view, introduction of heat to a gas is not the introduction of a substance, as many believed, but a *conversion* of mechanical work into kinetic energy of the molecules. This connection between work, heat and kinetic energy was tested by Joule in his 'churning' experiments. A thermoisolated substance was churned by a set of revolving blades for a long period of time. Even though no substance was introduced into the system, the temperature of the system increased by a small but noticeable amount. The success of these churning experiments not only served to vindicate the identity between temperature and mean kinetic energy, but also led to the downfall of fluid theories of heat.[7] Remember, this exciting prediction was based on an identity between the temperature and mean kinetic energy of a gas along with Leibniz's principle of the indiscernibility of identicals.

4. THE MIND–BODY IDENTITY THESIS

One of the most controversial applications of a reductionist philosophy has been the materialist's attempt to reduce the mental to the physical, making such claims as the science of psychology can be reduced to that of neurophysiology, that mental states are nothing but brain states if they are anything at all, and so on. (Some materialists even go so far as to say that mental state terms do not denote anything at all in that mental states are simply not real. This position is called eliminative materialism and it will be discussed below.) Whether the mental can be reduced to the physical, then, has been a perennial subject of much heated debate. In recent decades, this dispute has taken on several developments which indicate fresh approaches to the problem. Unfortunately, however, the waters may have been a bit muddied because each side of the controversy presents a picture of what it is to reduce one scientific system to another which does not quite do justice to the intricacies and complexities of actual, paradigm cases of reduction in the sciences. For example, many criticisms of materialistic viewpoints simply assume the D–N, traditional approach to the logic and nature of reduction. As a result, many demands are placed upon mind–body reduction which can not even be satisfied by paradigm cases of reduction between two physical sciences such as reducing classical thermodynamics to statistical mechanics. I believe that my model for reducing the mental to the physical should be based on the relatively non-controversial cases of reduction in the physical sciences. This model was developed in the preceding sections.

What does mind–body reduction look like according to my view? First of all, there is a reducing theory: any mental state, M_i, is identical with a neurological state, N_j. This is a programmatic claim in that our reducing theory presents us with a list of nominees for identity. We have yet to ascertain *which* N_j is the same as M_i, i.e., pinpoint our identities. Determining this concerns considerations that are partly analytical and partly empirical. By Leibniz's law, if $M_i = N_j$ then the two will be related in exactly the same way to all other mental and neurological states. The rest is 'easy'. Find a pair of M_i, N_j states which are *uniquely* related to *all other* states in exactly the same way and we will have the strongest available evidence for their identity. This is precisely what happened in the case of the identity of temperature and mean kinetic energy. Examining a list of empirically confirmed laws of each science, we notice that the mean kinetic energy of gas molecules is the only mechanical property which is related to the pressure and volume of a gas in exactly the same way temperature is. The nominees are thus narrowed down to a single mechanical property. The materialists have faith that someday identities between mental states and neurological ones will be pinpointed in like manner, knowing very well that the laws of psychology and neurology are, at present, neither as powerful nor as refined as those in the physical sciences.

One intriguing result of this version of materialism, one which has always been a source of temptation for the mind–body identity thesis, is that mental–physical causation ceases to be a mysterious phenomenon. Prior to this solution, mental phenomena were thought by some to be epiphenomenal, i.e., they were effects of physical causes but they had no power to affect the physical. Other philosophers wondered how non-material, mental phenomena could even be affected by 'hard' material phenomena. With the above version of mind–body reduction, the problem is easily settled. Since mental states are the same as neurological ones, by Leibniz's principle, whatever one causes, the other causes and whatever affects one, affects the other. So, for example, if a certain type of thinking is identified with a specific cerebral process and if we learned that such a process could be inhibited by a strong magnetic field, we could experimentally stop people from thinking in that way by placing them in the vicinity of a powerful magnet. Likewise, if the mental state of intending to raise one's arm is identical to a brain state which is causally connected with muscle contractions in the shoulder, it is easily understood how such a mental state can be, *is*, a part of a physiological, causal chain, ending with the raising of one's arm.

Another, very important result of the above account of mind–body reduction is that it can at least serve to clarify the nature of an age-old dispute between reduction and emergence. The latter view, which is the antithesis of reduction, maintains that the properties and processes of living things take place on an entirely different level than inanimate properties and processes; and, because of this, they are subject to macroscopic laws which are not reducible (or accountable) to the laws governing inanimate microprocesses. Interestingly enough, scientists from each side of the dispute claim that their view is the more scientific one in that the evidence is in their favour.

But what would count as evidence for reduction or for emergence? From what can be gathered in the literature so far, both sides of the dispute apparently have failed to realise how complex and delicate the issues of evidence are here. In order to show this, let us be as precise as possible in articulating what each side of the dispute claims. On the one hand, we know that the reductionist claims that because any macrostate will turn out to be the same as a microstate or a combination of microstates, both sets of properties take place on the same level and they obey the same laws. The emergencist explicitly denies this and insists that there exist an important class of macroproperties which are not identical with *any* microproperty or *any* possible combination of them. Hence, by Leibniz's principle, there are macroproperties which behave in a manner entirely different from any microproperty or combination of microproperties.

Now each side of the debate is committed to demonstrating something extremely difficult. Even if each list of macroscopic and microscopic properties (and all of their combinations) is finite, each position is confronted with a completeness problem. After all, does not the mind–body reductionist eventually have to show, among other things, that for any mental state M_i there is a brain state, N_j, that behaves in exactly the same way as M_i with respect to *all other* properties? In other words, the anti-reductionist need find only one aspect of M_i's behaviour that differs from that of N_j in order to refute the identity between them. But not so fast! Things aren't all that easy for the emergencist. Showing that a particular M_i is not the same as an N_j does not suffice to refute a reductionist thesis, for the reductionist may simply have claimed the wrong identity in the first place. The emergencist must show that there are mental states which are not identifiable with *any* possible combination of neurological states, which means that it must be shown that there are mental states whose behaviour differs from all possible combinations of neurological states. How can the emergencist systematically show this?

To make matters even worse, there is the *many-bodied problem*. In order to show that a particular macroscopic state behaves in a way other than any possible combination of microscopic states, one must know how extremely complex combinations of microscopic states behave in the first place. Wouldn't this include an understanding of the laws and models of interaction of these highly complicated systems under a variety of boundary conditions? Surely, the emergencist can not argue that because the behaviour of a many-bodied system does not in the least *resemble* the behaviour of its individual members we therefore must admit of the existence of new, emergent properties. A combination of magnetic fields may not in the least resemble an individual one, but so what? It certainly does not indicate special kinds of emerging fields. There may be scientific systems that deal with microscopic phenomena, which have not yet been sufficiently developed to tell us how highly complex combinations of microscopic entities behave; or, at least, we do not know what the theory says about this. (A situation very much like this existed for quantum field theory.) Determining how many bodies under a variety of conditions will behave can be a very deep problem. Yet, many philosophers who cite the strange and unusual behaviour of macroscopic systems in comparison to that of individual microscopic systems as evidence for emergence have simply failed to take into account that the weird behaviour of complex combinations of microscopic entities under a variety of conditions is the very thing we should expect according to the laws which govern many-bodied systems. So, until such laws are really understood, comparing the behaviour of an individual macrosystem with that of variously arranged microsystems has absolutely nothing to say of significance about the reduction-emergence issue. Of course, by the above, it does not mean that I am arguing in favour of the reductionist viewpoint over emergence. I am only trying to articulate what each side of the dispute is saying in precise terms in order to anticipate a resolution of the issues by empirical means.

5. PHILOSOPHICAL CRITICISMS OF THE MIND–BODY IDENTITY THESIS

We have seen how reductionist programmes are fundamentally based upon the application of the identity relationship to things and properties of the reduced and reducing sciences. The success of these programmes depends on establishing identities, and using them to explain and predict all kinds of relations between properties and things. Opponents of reduction, particularly mind–body reduction, have made a series of efforts to block the establishment of identities between mental and neurological states by piling restriction upon restriction on the very notion of a mental–physical state identity relation. It is to these demands I now turn.

The first demand I will discuss is the synonymy demand. It maintains that a necessary condition for property identity is that their corresponding predicates have the same meaning.[8] In other words, theory reduction takes place by systematically reducing the meanings of terms in the reduced science to those of the reducing science. So, it is argued, while temperature in thermodynamics is reduced or identified with mean kinetic energy in statistical mechanics because 'temperature' eventually comes to mean (or has the same use as) 'mean kinetic energy', it will never be the semantical course of scientific theory to identify the meanings of mental state terms with neurological state expressions. No matter what kind of neurological state may be referred to by 'I want an ice cream cone', it will never come to pass that such an expression *means* 'I am in brain state N_j.' Thus, mental states are not identifiable with neurological states because the language of the former can not be translated in terms of the latter.

Even if the above claim that mental state terms are not translatable into neurological state expressions is right, the demand that they be so translatable is wrong. Is the synonymy of predicate expressions a necessary condition for the identity of their corresponding properties? I think not. As Fodor,[9] Putnam[10] and Armstrong[11] have shown, there are countless cases of property identity where their corresponding predicate expressions differ in meaning. Returning to temperature of a gas being the same as the mean kinetic energy of its molecules, contrary to the above claim, 'temperature' does not mean 'mean kinetic energy'. The meaning of temperature is a function of classical thermodynamical theory and it is defined in terms of thermal equilibrium:

> The systems themselves, in these states, may be said to possess a property that ensures their being in thermal equilibrium with one another. We call this property *temperature*. The *temperature of a system is a property that determines whether or not a system is in thermal equilibrium with other systems.*[12]

In other words, when two systems are in thermal equilibrium, even though their pressures and volumes may differ, they have the same temperature. But it can also be shown that when two gases are in thermal equilibrium, their mean kinetic energies are equal. Except for the trivial case when their respective volumes and pressures are equal, mean kinetic energy is *the only mechanical property* of the same amount in systems in thermal equilibrium. What better argument could one give for the identity of temperature and mean kinetic energy? But this ratiocination has nothing to do with any synonymy between 'temperature' and 'mean kinetic energy'; on the contrary, the sole considerations are reference and the application of Leibniz's law to the behaviour of temperature and mean kinetic energy while the meanings of their respective terms are left to themselves. So, if in the paradigm case of the reduction of one scientific system to another, there is no indication whatsoever of the need to perform a translation of the terms of the reduced science to those of the reducing science, why should we demand that mental state terms be translated to neurological ones? Those who develop theories which reduce are scientists, not semanticists.

Another well-known criticism of the mind–body identity thesis is that it is a pseudo-scientific one in that the property identity claims contained within are not really supported by scientific evidence; they are really metaphysical assumptions in disguise. The major premiss of this critique is that, in terms of evidence, we can not distinguish between the identity of a mental state with a brain state, on the one hand, and their *correlation*, on the other.

Each hypothesis, it is maintained, generates the same set of predictions.[13] It seems that the only way we can test an identity claim is to observe correlations between states that are supposedly the same; but this is exactly how we would test for correlations between them. The only rationale, then, for opting for the stronger identity relation is to adopt an ontology that has fewer properties. But parsimony is a metaphysical principle, not a scientific one. Thus, the mind–body identity hypothesis is not really a scientific hypothesis at all.

I disagree with the above for several reasons. In the first place, the logic of each claim is such that they must lead to different predictions. An identity between a particular mental and neurological state automatically brings Leibniz's indiscernibility of identicals to bear while this is not the case for their correlation. Suppose an identity existed between mental state 10 and neurological state 25. We have: (1) $M_{10} = N_{25}$ implies for any other property, P, M_{10} will be related to P in exactly the same way that N_{25} is. However, because Leibniz's principle does not apply in the case of correlations, (2) M_{10} is correlated with N_{25} does *not* imply for any other property, P, M_{10} will be related to P in exactly the same way that N_{25} is. So, for example, if a particular mental process were to be identified with a brain process, and if the latter were discovered to be disrupted by sun spots, we could predict that people could not function mentally in certain ways whenever sun spots occurred. No such prediction follows from the mere correlation of these mental and brain processes. Let us take another example. Suppose we identified afterimages with bleached retinas. Such a claim entails that afterimages fade and bleached retinas diminish at precisely the same rate. Again, no such prediction follows from correlating afterimages with bleached retinas. In order to get such a prediction, *another* correlation must be added to the bleached retina–afterimage correlation, one which holds between the intensity of afterimages and the degree of bleaching. But whereas with the identity of bleached retinas and afterimages we have good reason to predict that afterimages will fade at exactly the same rate as bleached retinas diminish, what reason would we have to predict such a relation in case of a correlation, except that we have observed such a relation? – which is what we want to predict in the first place!

My reply to the above criticism should begin to look familiar to the reader. Not only does the identity between the mental and the physical lead to predictions other than those generated by correlations between the mental and the physical, the more correlations that are discovered to hold between mental and neurological states, the greater the evidence for identity in comparison to correlations. While it is true that a conjunction of correlations does the same predictive job as a single identity, we also know that because the correlations are independently maintained, they offer much less support for the *total* data than identity, as represented here:

$$\Pr(H/\text{Total } E_n) \frac{\theta_1 \qquad \theta_2 \text{ and } \theta_3 \text{ and } \theta_4 \ldots \text{ and } \theta_n}{.9 \qquad .9 \times .9 \times .9 \ldots \times .9},$$

where θ_1 stands for the identity of a mental and neurological state while θ_2, θ_3, θ_4, etc. are independently held correlations between these states. The total E_n, here, is the variety of different kinds of correlations found to hold between a particular mental and neurological state. What this means is that in the face of a variety of correlations between the mental and physical over various aspects, the identity hypothesis is the more confirmed.

Another objection to the mind–body identity thesis is in the form of a demand known as the *principle of simultaneous isomorphism*:[14]

If $M_i = N_j$ then an M_i type mental event occurs to a person X at a time t, if and only if a N_j type event occurs in the body of X at t.

This condition of the existence of a strict one-to-one correspondence between mental states and neurological ones presents the identity theorist with a difficult problem. Unlike the situations supposedly found in the physical sciences where it is easy or feasible to find correspondences between different *types* of physical properties, in the case of mind–body reduction, the discovery of type correspondences between mental and neural states is highly unlikely. The reasons given for this usually range from the belief that 'there are no firm data for any but the grossest correspondence'[15] to the claim that such type correspondences simply can not exist. The argument for the latter claim is based upon the contention that a plasticity among mental–physical states exists in a way that is not present in the physical sciences. To use several examples which can be found in science fiction, there may be cases where organisms of an entirely different neurological make-up may nevertheless realise the same mental state(s). In Hoyle's *Black Cloud*, a plausible case is made for an organism whose brain is simply a gaseous substance, albeit of an extremely complex organisation, to be in the same mental state as a human being while it is perfectly clear that the Black Cloud does not share in any of our neurological states. Therefore, possibilities such as these, along with the principle of simultaneous isomorphism, rule out mind–body identities.

The problem with the above critique is that it assumes incorrectly that type correspondence or the principle of simultaneous isomorphism holds in the natural science cases in the first place. It is important for us to see why this is not the case; in doing so, we will gain some new insights on what reduction is all about.

Contrary to what philosophers have preached about the by now overworked paradigm of reduction of classical thermodynamics to statistical mechanics, temperature is not strictly the same as mean kinetic energy; nor does there exist a universal connection between temperature and some mechanical parameter(s). Rather, *in an ideal gas*,[16] temperature and mean kinetic energy are the same, and we base this claim partly on the existence of a type correspondence between temperature and mean kinetic energy *relative* to an ideal gas. But no such correspondence can be found when we compare temperature and mechanical parameters in *other* situations. The actual state of affairs in thermodynamics is that we get a *variety* of relations between temperature and mechanical parameters, depending on whether we are dealing with ideal gases, solids, solids composed of distinguishable or indistinguishable particles, oscillators that behave continuously or discontinuously, and so on, i.e., the actual lawlike relation between temperature and mechanical parameters will be, as expected, a function of the ontology of the system in the manner described in the last chapter. The reducing theory provides a common ontological zoo of ideal gases, solids, Bose–Einstein solids, harmonic oscillators, etc. What mechanical property temperature corresponds to, then, will depend on what 'animal' we are dealing with in our ontological zoo. This is just applying ontological contextualism to reduction. This situation obviously violates the above type correspondence requirement in that temperature corresponds to a string of disjuncts if it corresponds to anything at all. Yet, for an ideal gas, temperature is the same as mean kinetic energy; for a solid, it may be identified with some other mechanical property. These identities hold even if we can not establish a universal one-to-one correspondence between temperature and a single mechanical property. Now, opponents can not argue that temperature does not thus exist or 'temperature' loses its meaning. Temperature is still that property of the system which determines its thermodynamic equilibrium, as it always was, i.e., its meaning is independent of any of these metaphysical models. Yet a physicist would surely insist on reducing macroscopic systems to configurations of microscopic entities and use changes in mechanical parameters causally to explain thermodynamical changes and vice versa.

Actually, requiring a universal correspondence between temperature and mean kinetic energy rests on a tempting mistake. It assumes that if there exists a single equation or relation that covers a wide range of phenomena, along with the same parameters fed into the equation, the relations between these variables ought to remain constant, thus entailing correspondence. But, of course, this neglects the fact that even though the equation and the parameters remain the same the *boundary conditions* may vary over the circumstances, yielding a variety of models and relations among the parameters. So, for example, the Schrödinger equation holds for a variety of physical circumstances but different sets of solutions under various boundary conditions will reflect different physical systems, e.g., the harmonic and anharmonic solutions of the equation.

If the principle of simultaneous isomorphism does not even hold in the thermodynamical–statistical mechanical case, it becomes unreasonable to require it in the case of psycho–physical reduction. In the same way that ideal gases do not work as solids do, why should a Martian brain or Hoyle's Black Cloud work in the same way as our brain? In other words, we should not even expect universal correspondence between the mental and the physical, for the same 'contextualist' situation may arise here; what identity we maintain to hold between a mental state type and neurological state type will depend on our models and theories of sentience in conjunction with other theories of how (biological) systems work. Naturally, there is no state of the art today, but its theoretical feasibility can not be ruled out by proving the absence of simultaneous isomorphism.

What I have been advocating all along has been the replacement of absolute identities, i.e., identities that hold no matter what ontology the scientific situation may call for, with *relativised* or *contextual* identities, i.e., the identity maintained *depends* on the ontological context, as it was in the case of reducing temperature to mechanical parameters.

Although the above move frees the reductionist from the demands of universal correspondence and absolute identities, making it is not so innocuous as it may appear. For one thing, if we are to preserve the transitivity of identity, the property, *temperature of a gas* can not be identical to the property, *temperature of a solid* simply because each, in its own context, is identical to *different* mechanical properties. In other words, even though 'temperature' designates that property of a gas, solid, etc., which determines its thermoequilibrium – 'temperature' in 'the temperature of the gas is 85 F°' being synonymous with 'temperature' in 'the temperature of this solid is 85 F°' – temperature observables are 'split', i.e., the temperature of a gas is not the same as the temperature of a solid, and so on; in fact, what temperature we ascribe to an object will depend, in part, on what *other* properties that object has, whether it is a solid, liquid, gas and so on. This is what it means to say that temperature is a contextual property and that an identity between temperature and some mechanical property is relativised to the ontological context. If such a move disturbs our philosophical common sense, we should try to realise how natural it becomes when we take up actual working situations in the sciences, that contextual identities are an essential feature of reduction.

6. FUNCTIONALISM

Functionalism essentially accepts the principle of a simultaneous isomorphism critique of the identity thesis.[17] The functionalist believes that because mental states can be realised in an infinite variety of ways, in a vast variety of physical systems, we can not find a non-trivial physical property *in common* to all these realisations to warrant any identity between mental and physical states. To use a well-known example of Putnam's, a pile of sawdust may be able to process the same information as a human being provided it possesses the proper complexity. What the functionalist does is to reduce mental states to *Turing states*, the latter being abstract or computational states that mediate between computer (= Turing machine) inputs and outputs. (Cf. the discussion of computer simulation models in Chapter 3.) Functionalism is based upon an analogy between minds and computers. Crudely, thinking is a kind of data processing and human beings are to be thought of as organic computers. Turing states´are computational states, i.e., they are abstract states or steps a computer must pass through in order to process information. A Turing state's identity, so to speak, is determined by its location within the functional organisation of the Turing machine given by the machine table. For example, *preferring* is that Turing state 'which controls the behavior of the machine (more precisely, the function which together with the machine's inductive logic controls the behavior of the machine) assigns a higher weight to the first alternative than the second'.[18]

What this means is that while a human being, a Martian, Hoyle's Black Cloud, *et al.* will not have any physical property in common, they all share the same Turing or computational state when thinking a certain thing. In other words, their psychological life is reduced to that of an abstract system of data processing. The laws of psychology are really laws which govern how information is processed. But these systems are abstract in that their ascription is not dependent on material make-up. Whether a particular programme of computational states is to be ascribed to a computational device depends on how well it can explain the output of the device as a function of input. This is what psychology is all about. As psychologists, we are trying to guess how human beings are programmed, and we can do this without knowing whether the human brain is made out of protoplasm or sawdust. Thus, because functionalism identifies mental states with abstract, Turing or computational states *vis-à-vis* physical states, it is perfectly compatible with mind–body dualism.

My version of mind–body reduction does share this with functionalism: mental states are to be functionally defined in terms of their analogy with Turing machines. In the same way that 'temperature' is functionally defined as that state which determines thermoequilibrium, a mental state will be defined in terms of its computational role within an entire computational system.

A mental state would be part of an internal, computational sequence between input and output in exactly the same way that thoughts mediate between stimuli and behaviour. Thus mental states are functionally defined as those computational states *which determine* the behaviour of the individual organism under various inputs.

Unlike the functionalist, however, I have argued that mental states (or Turing states) are identifiable with physical states *even if there does not exist a non-trivial physical property in common to all realisations of mental states*. This is what contextualist identities are all about. In the same way that what particular mechanical property temperature is identified with depends on which member of our ontological zoo we are dealing with, the material reference of a particular mental state depends on the physical nature of the information processing system in question, whether it is a human brain, a Martian or a Black Cloud. Just because these information processing systems do not physically work in exactly the same way, it does not mean that a mental state is not identifiable with a physical state.[19] It just means, as I have said, that such an identity is qualified by the ontological context.

In fact, I insist that the view of mind–body reduction by contextual or relative identities makes more sense of what the functionalist claims that the functionalist who disavows mental–physical state identities. If mental states are reduced to Turing or computational states, according to the functionalist, how then are Turing states related to physical states? The functionalist's answer is that the latter are (physical) realisations or instantiations of the former. But then *how* can a physical state or an information processing system be considered to be a realisation of a mental or Turing state in the first place? While it seems that the functionalist story can not make good sense of the notion of physically realising a mental state, I believe that my view of contextual identities does. There is no reason whatsoever to ascribe temperature to a collection of molecules except to say that a collection of molecules *has a mechanical property that determines its thermoequilibrium* (which happens to be mean kinetic energy in the ideal gases case). How else could we say that temperature has a mechanical realisation? By analogy, how can mental states (or Turing states) be said to be physically realised except that a certain physical system has a physical state which functions in determining its behaviour in exactly the same way as a mental state, i.e., they are the same states within this context? Thus, the notion of a physical realisation of a mental system is parasitic upon the identity relation, albeit a contextualised identity. I see no other way that such realisations can take place.

7. ELIMINATIVE MATERIALISM

Eliminative materialism claims that the reduction of the mental to the physical ultimately amounts to showing that mental state terms do not refer. The fate of mental predicates is analogous to past theoretical expressions such as 'phlogiston', 'the aether', etc., where they were once thought to have a legitimate reference but it was subsequently shown to be illusory.[20] Now an interesting problem arises with my picture of reduction. With contextual identities, it turns out that although the macroscopic reference of 'temperature' is fixed, when the macroscopic systems of classical thermodynamics are reduced to the microscopic systems in statistical mechanics, the microscopic reference of 'temperature' *varies* from one context to another. Does this feature liken my view of reduction to that of the eliminative materialist? One might answer this affirmatively, pointing out that before reduction 'temperature' was thought to refer to a single property, that in so far as it now refers to many things, it does not denote anything real.

We must be very careful here. Even if the microscopic reference of 'temperature' is non-uniform we must remain cognisant that the reference of 'temperature' is fixed, in part by being entrenched in classical thermodynamics in the same way that 'phlogiston' was entrenched in a past chemical theory.

Suppose we now insist that temperature is not a real property because of non-uniform microscopic reference. Surely, this case is not at all like the discovery of chemists that phlogiston was not real? In the phlogiston case, it was shown that nothing answerable to the properties and relations of phlogiston was responsible for combustion. An entire chemical theory was eliminated along with it. Unlike this case, the reduction of classical thermodynamics reinforces our belief in classical thermodynamics, for we not only believe that temperature is that property responsible for thermo-equilibrium, we now know what the temperature of gas, solid, etc. *is*. Unlike the phlogiston example, we have not rejected the (macroscopic) relations of classical thermodynamics, in particular, state descriptions of these systems. I feel that a reduction which eliminated temperature as a property must bring down classical thermodynamics as well; eliminative materialism requires a similar downfall of psychological systems.[21] There is a very simple way to bring this out. According to my view, we reduce thermodynamical states to microphysical ones by establishing identities between them. But, by Leibniz's principle, if temperature is the same as mean kinetic energy, then temperature is just as real as mean kinetic energy. Likewise, if mental states turn out to be the same as neurological states, the former are just as real as the latter. This form of reduction, then, is a far cry from elimination of the reduced system of properties.

It is true that the proponents of classical thermodynamics *may* have taken temperature to be a simple *mechanical* property. Perhaps the fact that it is not might lead one to insist that temperature is not real. There are problems raised by this that can not be adequately dealt with here. They concern realism and uniform reference. Should we be tempted to adopt this principle concerning the reality of a property?

U: A necessary condition for the reality of *P* is that *P*'s reference to microscopic properties be uniform upon reduction.

Although *U* appears to be an interesting and powerful ontological principle, I doubt that it would be useful to adopt it. Even if temperature *uberhaupt* is not real according to *U*, what about the temperature of a gas, solid, etc.? Is not the temperature of a gas related to its pressure and volume? Do not temperature and mean kinetic energy behave exactly the same way in an ideal gas? Does not an increase in the temperature of a gas *cause* (when the volume is held constant) its pressure to go up? Again, it can be seen how this contrasts with the phlogiston case. It appears that we can handle these cases without adopting *U*. Our model of reduction tells us that what may have been taken on the macroscopic level to be a single property is many on the microscopic level. So we flag '*P*' whenever we use it. But isn't this what being a contextual property is all about? In terms of the reality of *P*, does it make any difference so long as we make such qualifications on our use of '*P*' clear?

It is most important to make clear that even though reference loses its uniformity in the case of reduction by contextual identities, this has no effect whatsoever on the model of reduction presented in the earlier sections. It is still a case of ontological reduction. Reduction still takes place by identifying each and every entity and property of the reduced science with the entities and properties of the reducing science. Leibniz's principle is still that principle which guides us in pinpointing these identities. Identities comprise the major content of reductionistic explanations of macroscopic phenomena. The only difference is that the reference of some of the macroscopic *terms* is non-uniform. Even so, even with non-uniform reference, the total number of properties – which equals the number of macroscopic plus the number of microscopic properties – is reduced because more microscopic properties are captured by the identity relation as we go from one context to another. In other words, even though our identities are contextual in nature, we are still reducing the number of independent phenomena by identifying them with the very same aspects of a common ontology provided by the reducing theory.

The moral of my story is this. If a successful mind–body reduction ever takes place, it will not in the least resemble reduction as it has been traditionally characterised in the philosophy of science. The traditional approaches to mind–body reduction (and the mind–body identity thesis) have conspicuously left out the notion of reducing the number of entities and properties there are in the universe by identifying them with the reducing theory's system of objects, and they have completely overlooked the possibility of establishing identities which are not absolute but relativised to the nature of the entity selected for investigation by the scientist. I realise how heretical these views are but I think that they will serve as the best possible guide for appreciating future developments concerning the unification of the behavioural and natural sciences.

NOTES

1. See Hempel, *Philosophy of Natural Science*, pp. 101–10.
2. Much of the research in sections 1–3 of this chapter is based on an article by T. Ager, J. Aronson and R. Weingard, 'Are Bridge Laws Really Necessary?', *Noûs*, vol. VIII, no. 2 (1974).
3. Hempel, *Philosophy of Natural Science*, pp. 72–5.
4. Cf. T. Ager *et al.*, 'Are Bridge Laws Really Necessary?', pp. 119–22.
5. Ibid., pp. 122–5.

6. I owe this point to R. Weingard.
7. T. Ager *et al.*, 'Are Bridge Laws Really Necessary?', pp. 124–5.
8. Cf. W. Sellars's 'The Identity Approach to the Mind–Body Problem', in *Philosophy of Mind*, ed. S. Hampshire (New York: Harper & Row, 1966) pp. 7–30.
9. J. Fodor, *Psychological Explanation* (New York: Random House, 1968) pp. 100–6.
10. H. Putnam, 'Minds and Machines', in *Minds and Machines*, ed. A. Anderson (Englewood Cliffs: Prentice-Hall, 1964) pp. 85–94.
11. D. Armstrong, *A Materialist Theory of Mind* (London: Routledge & Kegan Paul, 1968).
12. M. Zemansky, *Heat and Thermodynamics* (New York: McGraw-Hill, 1957) p. 7.
13. R. Brandt, 'Doubts About the Identity Theory', in Sidney Hook (ed), *Dimensions of Mind* (New York: Collier Books, 1960).
14. R. Brandt and J. Kim, 'The Logic of the Identity Theory', *Journal of Philosophy*, vol. 64 (1967) pp. 515–37.
15. J. Fodor, *The Language of Thought* (New York: Crowell, 1975) p. 17.
16. I owe this point to R. Weingard.
17. Proponents of functionalism have included J. Fodor and H. Putnam.
18. H. Putnam, *Philosophical Papers*, vol. 2 (Cambridge University Press, 1975) p. 410.
19. What this means is that there are Brandt–Kim correspondences or bridge laws between mental and physical states but each one must be qualified by its particular context. Likewise while a Martian 'mental' state may be the same Turing state as a human mental state, they may refer to different microphysical states.
20. See R. Rorty's 'Mind–Body Identity, Privacy and Categories', in S. Hampshire, *Philosophy of Mind*, pp. 30–63.
21. See B. Enc's 'Reference of Theoretical Terms', *Noûs*, vol. x (1976) pp. 261–82.

9 Counterfactuals

Counterfactuals (or contrary-to-fact conditionals) are propositions of the form, 'If X *were* the case then Y would be the case', 'Even though X did not occur, were it to have occurred, Y would have occurred', 'If there were an X, then there would be a Y', and so on. Every science is inundated with claims of a counterfactual nature, even history, yet, while such claims are made all the time in science and in everyday life, they have notoriously resisted traditional logical tools of analysis. Recall the problems they presented Hempel's D–N model. If the ' $\supset$ ' or material implication is used to capture the logic of a counterfactual, i.e., if counterfactuals are a special kind of conditional, then intractable problems arise because any and all counterfactuals, as well as their contraries, would be equally true. This absurd result follows simply from the definition of material implication and that the antecedent of a contrary-to-fact conditional is not realised. (Otherwise, it would not be contrary-to-fact.) Again, if we use the ' $\supset$ ' to capture the logic of counterfactuals then, in the case where I don't drink a glass of water which is placed in front of me, 'If I were to drink it, my thirst would be quenched' and 'If I were to drink it, I would become deathly ill and my thirst would not be quenched' are *equally true*, which is absurd. It is my contention that our theories tell us which counterfactuals are true and that counterfactual support is a partial function of a theory's ontological commitments, no matter how sophisticated or naive our theories may happen to be. In other words, the very same features of a theory that led to explanation and the fixing of probability measures for events are also responsible for counterfactual support.

What we are seeking is a means of understanding how it is, for example, that Newtonian mechanics supports 'Although this particle is moving at 99 per cent of the speed of light, were we to increase its kinetic energy, it would move faster than light' while the special theory of relativity does not. Or, in psychology: one psychological theory might support 'If this individual were placed in a more nurturing environment, he would score significantly higher on an IQ test' whereas such a claim would not be lent support from another psychological point of view. So, we wish to come up with a criterion which tells us if a counterfactual in question is supported by a particular theory.

1. THE CONSEQUENCE APPROACH TO COUNTER-FACTUALS

The most natural way logicians have come to grips with the failure of truth functional analyses of counterfactuals has been to base the truth of a counterfactual claim in terms of its derivability from laws and initial conditions instead of the separate truth values of the antecedent or consequent of 'If X were the case, Y would have been.' So, counterfactual claims are conditionals that logically follow from assumed initial conditions and laws. Thus, 'If I were to release this chalk, it would move towards the floor' is true *relative* to Newtonian mechanics but not 'If I were to release this chalk it would move into orbit' because the former follows from Newton's laws under these conditions while the latter does not. Notice, they both have the same truth functional status because, in each case, the chalk is actually not released. Nevertheless the former is true while the latter is not because the former follows from Newton's laws in combination with the facts while the latter does not. It appears that the above problem of counterfactuals has been neatly solved.

What, then, does the consequence analysis of counterfactuals have to do with the ontology of a theory? How does a theory's ability to reduce the number of independent phenomena lead to counterfactual support? While I agree with the overall approach of the consequence view, I feel that it has shortcomings, especially because it places the entire burden of counterfactual analysis on the derivational features of propositions rather than the ontological features of the actual situation which serves as a base of counterfactual support. For example, in the above glass of water case, I want to say that the water *would* quench my thirst and that it is not physically or medically possible for it to make me ill were I to drink it because of the intrinsic properties of the water. It simply does not have the stuff to do it, whereas water laced with arsenic does.

Counterfactual claims are about possibilities but they involve actualities as well. And this is where ontology comes in. According to my view, the possibilities involving the behaviour of things are a function of their nature, their intrinsic properties. What counterfactuals hold true of a system will depend on the ontology of the situation, the way things are, the way they are arranged, etc. Possibilities stem directly from the actual state of affairs. (Such a position has been called combinatorialism because possibilities are seen as various combinations of properties of things in accordance with the laws of nature.)

So, we are left with this general version of the consequence analysis of counterfactuals. Suppose we have a system and we wish to know if a particular counterfactual holds of that system. In order to answer this question, we must first come up with a complete catalogue of the intrinsic features of the system. Once this is done, we are then in position to know which laws of nature apply in order to determine if the counterfactual in question is a logical result of said features and laws. For example, we want to know if 'Even though the chalk is not released, if it were, it would accelerate towards the ground' is true. The cursory answer is: since the chalk is endowed with mass, immersed in a gravitational field, etc., we know that Newton's laws are applicable to this system; it then follows that the chalk will fall when released. Thus, even though the chalk is not actually (or even ever) released, Newtonian mechanics tells us that were it to be released, it would fall.

While the above sketchy analysis of counterfactual support is to be considered a version of the traditional consequence approach, there are differences as well. I will just mention two salient ones for now. In the first place, the traditional analysis does not properly distinguish between the ontology of the counterfactual situation prior to the occurrence of the antecedent and those features which are related in the laws of nature once the antecedent obtains. Secondly, the laws themselves are provided a truth functional analysis which, I believe, is a grave mistake for reasons similar to those already mentioned in Chapter 3. As a result, the consequence view of counterfactuals has been open to a singularly powerful criticism, known as the problem of cotenability; so powerful has this attack been, it convinced several philosophers to give up the consequence approach altogether and seek out others. So the view that the ontology of a theory in combination with its laws it what is responsible for counterfactual support must come to grips with this highly difficult problem, which is what I will cover in the ensuing sections of this chapter. However, the reader is more or less warned that handling the cotenability problem requires that we traverse many technical labyrinths which may or may not be worth the effort for some, depending, of course, on one's background (especially, in symbolic logic) and interests. If the reader decides that he or she does not have the prerequisites, it is advisable to skip to the last section of this chapter.

2. THE COTENABILITY PROBLEM

The problem of cotenability was first introduced by Nelson Goodman in *Fact, Fiction and Forecast* over twenty years ago. Although several solutions to this puzzle have been proffered by logicians and philosophers of science, the critics of the consequence view have remained unsatisfied, as well as many of those who nevertheless prefer the consequence analysis over its rivals. Below, I will add my attempt to crack this mystery, I hope in such a way as to add some insights on the nature of counterfactuals. I have suggested above that the problem of cotenability rests on some mistaken assumptions about the nature of counterfactuals and the laws which lead to their support. In relation to this, another source of confusion arises from the improper use or application of laws (or equations) to specific situations, which results from the improper substitution of conditions into lawlike expressions. Specifically, I will argue that, contrary to the orthodox view, counterfactuals are *not* conditionals in spite of their grammatical appearance; that straightforward universal conditionals do not capture the logic of laws of nature; and, that the problem of cotenability results, in part, because of the mistaken belief that the derivation of a counterfactual from laws and conditions requires that the counterfactual's antecedent be thought of as a completing condition of a set of sufficient conditions for its consequent. Finally, I will argue that the original formulation of the cotenability requirement is misleading, that it is so characterised as to go beyond consistency demands, and in such a way as to entail what I believe to be an erroneous analysis of counterfactuals.

Again, the consequence view holds that counterfactuals are conditionals that logically follow from assumed initial conditions and laws. To use a venerable but worn-out example, 'If this match were struck it would light', is unpacked:

1. $(O . D . M . S) \equiv L$ (or $(O . D . M . S) \supset L$) Law
2. $O . D . M . (\sim S . \sim L)$ conditions
$/ \therefore S \supset L$

O – The match is in the presence of oxygen.
D – The match is dry.
M – The match consists of combustible material.
S – The match is struck.
L – The match lights.

I place parentheses around $\sim S . \sim L$ because although they are responsible for the contrary-to-fact nature of the conditional, they are nevertheless not supposed to partake in the derivation of $S \supset L$. But, as Goodman realised, their truth is troublesome, for we can also get $S \supset \sim (O . D . M)$ from the very same premises, *no matter what their content or what our actual laws of nature may happen to be.*[1] The reason for this is that we can always construct a conditional proof, starting out with S, and using $\sim L$ in 2 to get $\sim (O . D . M . S)$ via *modus tollens*. We can then use combinations of O, D and M to derive $\sim O$, $\sim D$ and $\sim M$. Hence, the consequence analysis of counterfactuals has the difficulty of deducing the prerequisite conditional in such a way as to exclude a rival conditional whose antecedent implies the denial of the original conditions. In the above example, we want to derive 'If the match is struck it will light' from our laws and conditions and we also want to deny the truth of 'If the match is struck it will not be dry.' From the above, however, there seems to be no *logical* reason for denying $S \supset \sim (O . D . M)$. Goodman concludes that establishing the compatibility of S with our initial conditions actually rests on the truth of *another* counterfactual: $\sim (S \supset \sim D)$. Hence, the consequence analysis of counterfactuals appears to be doomed to circularity by explaining counterfactuals in terms of other counterfactuals.[2]

Now, right away, the natural thing for the consequence theorist to do is to cry 'foul', for $S \supset \sim (O . D . M)$ followed only through the use of $\sim L$ as an initial condition, and that it is unfair to use contrary-to-fact conditions $(\sim S . \sim L)$ in *any* derivation of a conditional. 'Even so', our cotenability critic will reply, '$\sim S . \sim L$ are nevertheless *true*, and no amount of ratiocination will make their truth go away'. I agree with this, and it reveals a major shortcoming with the above version of the consequence analysis, as well as one crux of the cotenability problem: a natural, logical motivation must be supplied to bar $\sim S$ and $\sim L$ from serving as initial conditions in our analysis. As things stand, it is not there.

There is another, more general version of the problem. The strategy of the consequence model is to catalogue enough conditions about the system in question such that when the antecedent of the counterfactual is added to this list, derivation of the consequent ensues; and in such a way that the antecedent does not lead to the denial of the initial conditions. The problem with this procedure is that we have no idea about any far-reaching consequences of lengthening our initial conditions.[3] For example, suppose in the above match case that a 'moistening device' is attached to the match in such a way that the match, while dry before it is struck, activates the device when struck. It appears, here, at any rate, that 'If the match were struck, it would light' is false and 'If the match were struck, it would not be very dry' is true. In other words the conditions immediately before the occurrence of the antecedent and laws are such that the occurrence of the antecedent nullifies one of the conditions. We could always realistically construct cases where this results.

The obvious reply for the consequentialist is that we have simply added another condition, say, the 'moistening device', and we can straightforwardly derive the truth of 'If the match is struck, it will not be dry' instead of 'If the match is struck, it will light' by means of our laws and this *new set of conditions*. While such a reply may sound convincing, it does not bear closer scrutiny. This is because the logical form of the above consequence model of counterfactuals is simply too clumsy to incorporate these additional conditions. As a result, we have no logical way to decide which set of conditions and antecedents leads to what counterfactual.

Let us see how this works. If the consequentialist admits that the addition of new conditions can entirely change the complexion of counterfactual support – imagine the truth of 'If the match were struck, it would light' oscillating back and forth as we add each new condition – then he must be committed to the existence of a single set of *complete* conditions for each circumstance, one which decides between rival counterfactuals whether he, in fact, knows which set is the one that is, in fact, true. Whereas Lewis's possible worlds analysis most readily handles this type of situation, because he can have a possible world correspond to each new set of conditions,[4] I do not believe the above formulation of the consequence model can. This is because we will need a (new) law for each new complete set of conditions in order to account for the truth or falsity of the counterfactual, $A \supset C$:

$$[(C_1 \ldots C_n) . A] \supset C$$
$$[(C_1 \ldots C_n . C_{n+1}) . A] \supset \sim C$$
$$[(C_1 \ldots C_n . C_{n+1} . C_{n+2}) . A] \supset C$$
$$[(C_1 \ldots C_n . C_{n+1} . C_{n+2} . C_{n+3}) . A] \supset \sim C$$
$$[(C_1 \ldots C_{n+1} . C_{n+2} . C_{n+3} . C_{n+4}) . A] \supset C$$

and so on. For example, we might claim under the originally stated conditions that if the match were struck, it would light. Then, if we were to add the moistening device (C_{n+1}), striking the match would not imply that it lights but that it does not light; however, if a dehumidifying device were, in turn, attached to the moistening device (C_{n+2}), striking would imply lighting, and so forth. It seems that without such an array of laws which connect these sets of conditions plus the antecedent of the counterfactual with its consequent, the above derivations could not be made. Without them, we would not have the slightest idea if conjoining the antecedent of the counterfactual in question with a given set of conditions would actually entail that the denial of at least one of the initial conditions required to derive C, a result that would lead to the opposite consequence or, at least, not allow us to derive C.

It is clear, however, that there are no such laws, that is, *in the above form*. Herein lies one major defect with the traditional version of the consequence theory: laws are expressed as truth functional conditionals (or biconditionals) of the form $[C_1 \ldots C_n) . A] \supset C$ (where '$C_1 \ldots C_n$', 'A' and 'C' respectively stand for conditions, the antecedent and consequent of the counterfactual). They are simply too static when expressed in this way. In fact, it is my contention that these expressions are not even laws. One reason why I maintain this is that they are not *adverbial* in nature. What I mean by this is that one salient feature about laws is that they do not describe what happens when certain conditions occur in nature but laws are extremely sensitive about how they occur. And, I believe paying cognizance to this very feature about laws is the key to the solution of the cotenability problem. We should not conclude that the match will light if it were struck without paying attention to *how* it was struck. After all, what if it were struck so rapidly that a partial vacuum resulted, removing all of the oxygen from the vicinity of the match? Any characterisation of a law, then, must capture these adverbial features. This aspect about laws becomes most apparent when we notice their logic in the form of equations in physics. So let us now examine some simple cases of counterfactuals in physics and see if they offer us any insights into the cotenability problem and the nature of counterfactuals.

3. COUNTERFACTUALS IN PHYSICS

Suppose I am holding a body, *a*, above the floor. Given that such body is endowed with mass, that it is located in a gravitational field, that it is so many feet above the floor, etc., what counterfactuals can be inferred if I were to release it? The answer, I submit, is it depends, among other things, on *how* I were to release *a*. In fact, the best answer appears to be that an indefinite number of counterfactuals would follow:

> If I were simply to release *a*, it would accelerate towards the floor at 32ft/sec².
> If I were to throw *a* up as I released it, it would not fall but move away from the floor.
> If I were to throw *a* out as I released it, it would describe a parabolic arch.

All of these would be true given that *a* is a Newtonian mass suspended above the floor. They *all* follow from Newton's law of gravity, along with his laws of motion, in particular, his second law, Force = mass × acceleration ($F = ma$).

The fact that a single set of conditions and laws can support a vast array of counterfactuals holding between various antecedents and consequents has apparently been overlooked in the literature. It is usually an either/or proposition for a counterfactual or its contrary (or neither) for each set of conditions and laws. This result follows directly from the traditional analysis. Given the actual circumstances and laws, as they are usually expressed, 'If the match were struck it would light' or 'If the match were struck it would not light' seem to be our only contenders but never 'If the match were struck very slowly, it would not light', 'If the match were struck *very rapidly* it would light', 'If the match were struck with a *tremendous amount of pressure* it would not light, (but break)', 'If the match were struck in a circular motion it would still light', etc. Yet, why can't these latter *all* be true under the same circumstances and laws? I maintain that they can but the traditional version of the consequence has difficulties with handling this feature, again, because of the ways they treat laws. The only natural way, it seems, for the traditional version of the consequence view to accommodate this feature about counterfactuals in physics is to depict the law(s) used to derive this set of conditionals as a conditional the antecedent of which states the actual circumstances $(C'_1 \ldots C'_n)$ while the consequent consists of a string of conjuncts, each conjunct being one of the derived conditionals $(A_n \supset C_n)$.

1. $(C_1' \ldots C_n') \supset [(A_1 \supset C_1) . (A_2 \supset C_2) \ldots (A_n \supset C_n)]$
2. $C_1' \ldots C_n' \quad \therefore [(A_1 \supset C_1) . (A_2 \supset C_2) \ldots (A_n \supset C_n)]$.

So, instead of having a new law for each new set of conditions, as it was in the last section, we now have a proposition as expressed in 1. While I believe this version is a considerable improvement over the original version, the problem of cotenability unfortunately remains, for $\sim C_1 . \sim C_2 \ldots \sim C_n$ can always be added to 2, yielding conditionals such as $A_1 \supset \sim (C'_1 \ldots C'_n), A_2 \supset \sim (C'_1 \ldots C'_n), A_3 \supset \sim (C'_1 \ldots C'_n)$ and so on. So, we are back at square one.

There is another, even more crucial problem with propositions like 1. 1's consequence consists of an infinite string of conditionals, and it behooves us to ask how such a string can be *generated* in a non-*ad hoc* manner. The only way out of this is to say that our laws of nature generate this string of conditionals but this means that *laws of nature contain more information than propositions like 1*. Then 1 cannot adequately express a law of nature but, instead, something that *follows* from a law(s). Hence, the traditional analysis requires even more revision.

I will begin this revision of the traditional model of the

consequence view by making a few remarks about logically representing laws in physics. I mentioned before that laws in physics were adverbial in the sense that not only did they cover those situations where conditions are satisfied but how they were satisfied as well. This requires a minor adjustment to the usual logical characterisation of laws. Instead of speaking of laws as universal conditionals which usually play a fundamental role within a scientific deductive framework, I want to say that lawlike expressions must contain invariant relationships between properties.[5] Thus, 'All ravens are black' is not a law, according to my view, because the only relationship occurring in that expression is a truth functional one between propositions. However, 'If x is a raven then there exists such and such a relationship between its genetic structure and the pigmentation of its feathers' comes a lot closer to being a law. Likewise, Einstein's famous law that nothing can travel faster than the speed of light would look something like this: for any x if x is endowed with mass–energy then the velocity of x is either less than or equal to the velocity of light (in a vacuum). There are then two parts of these lawlike expressions. The antecedent refers to a set of conditions which characterise the essential properties or features of the system in question. If a system meets these *ontological conditions* then the invariant relationship depicted in the consequent applies to that system. Likewise, the invariant relationship is naturally adverbial in nature because the values of the variables depict not just what things occur to the system in question but *how* they occur as well. For example, the relationship in Newton's second law, $F = m\,a$, not only tells us that forces bring about acceleration, but how much acceleration results from so much force, i.e., how bodies are accelerated. Each value of a variable, then, is an adverbial qualifier on the type of action occurring to the system.[6]

The ontological conditions for the application of the invariant relation in question are relativised to the counterfactual situation and our theories about how things work. Take, for example, a Newtonian mass which is suspended in a gravitational field and which also happens to be blue in colour. Now, our theories tell us that its blueness is not a relevant condition *vis-à-vis* the body's dynamical behaviour because whether Newton's laws apply here is *independent* of colour. On the other hand, a good case can be made that the body's colour is a relevant condition for relationships depicting how it would appear under various visual circumstances (e.g., it would appear purple under a red light) while its mass would not be a relevant consideration.

The distinction between the preconditions for the application of an invariant relationship and the adverbial qualifiers of the relationship, itself, is not just a logical but an ontological one as well; in fact, it is one which most naturally reflects the contrary-to-fact nature of counterfactual claims. As we already know, the truth of any counterfactual must be supported, in part, by the facts or the actual state of affairs. This feature is covered by the preconditions. However, we also know that even if we have enough facts at our disposal to inform us that a particular invariant relationship holds, it does not follow that anything has actually occurred to the system which is an exemplification of one of the adverbial qualifiers. For example, suppose we have a Newtonian mass which is, as a matter of fact, at rest. These facts alone, namely that the body is at rest and has inertia, suffice to allow for the application of Newton's second law, even if no forces whatsoever are being or ever will be exerted on it. This is nice because this feature captures what makes a counterfactual contrary-to-fact. So while the preconditions constitute those facts, along with invariant relationships, which make our counterfactuals true, the occurrence of any event represented by an adverbial qualifier of the invariant relationship need not ever take place; and, since the same set of conditions and laws may support an infinite number of counterfactuals, it means that *there is no factual difference among them*. In other words, all of these counterfactuals are on equal par when it comes to the facts. This cannot be overemphasised, and I will come back to this point again when my proposed analysis is contrasted to yet another aspect of how counterfactuals are usually viewed.

The above distinction between ontological conditions and the invariant relationship in lawlike expressions can be viewed yet another way. Since counterfactual information concerns what would *happen* to a system were the antecedent to obtain given the actual facts (even though the antecedent hasn't obtained), we are ultimately seeking information about what happens to the system as a result of the occurrence of this antecedent. So, the adverbial features of the invariant relationship usually take on an event status. On the other hand, the preconditions do not denote an event which occurs to the system in question but the *state* of the system *prior* to anything happening to it. A system having, say, inertial mass, being located in a gravitational field, etc., is not something that happens to a system but something it has, whereas being released is something that happens to it. So, each adverbial qualification denotes an event which is an action(s) occurring to the system in question, one which leads to a modification of the system. More important, each member of the set of adverbial qualifiers *denotes the same type of event or action*, differing only in *how* the action takes place. For example, in Newton's second law, the independent variable is the amount of net force which brings about a specific acceleration, i.e., each member of the set of adverbial qualifiers stands for a net force but *how much* net force is represented by different adverbial qualifiers. (I should like to note, in passing, that there are laws where there is an exception to this. Laws in which the invariant relations are static in that the relationships are simply between states are the exception to the above. For example, the Boyle–Charles law, $PV = nkT$ (where 'P', 'V' and 'T' stand for the pressure, volume and temperature of an ideal gas) is adverbial but pressure, volume and temperature are properties of the gas *vis-à-vis* things that occur to the gas. Nevertheless, any such static law can yield counterfactual information about changes that occur to the system simply by differentiating with respect to an independent variable. So, from the Boyle–Charles law and the fact that the system in question is an ideal gas, we can support 'If the temperature of the gas were to be increased (volume remains constant) its pressure would go up.')

In physics, the invariable relationship is expressed by an equation of some sort, where we have a functional relationship among variables of the form $x = f(y, z, q, r, \ldots)$, such as $F = ma$ (Newton's second law), $S = 1/2gt^2 + V_0t$ (the law of freefall), $A_1 = T_1A_2/T_2$ (Kepler's law of equal areas), and so on. The variable on the left is referred to as a dependent variable and it is a function of independent variables. In mathematical parlance, when we say that a dependent variable is a function of a set of independent variables, we mean that there exists a mapping from one set to another, where for each element in one set (the domain), we assign a unique element in the other (the range). The point in the range that is hit by an element, Xn, in the domain is known as the *image of xn*, $f(Xn)$. So the relationship between the variables in our equation is that of a mapping or correspondence, f, between sets of numbers. The invariant relationship within lawlike expressions, which is represented by this mapping, is what carries the burden of lawhood; and it is in contrast to the traditional analysis which instead places the burden on the truth functional relationship of material implication. It should also be emphasised that my addendum to the traditional analysis of laws is not at all intended as a full-blown analysis of lawhood; rather, I am suggesting that if laws were so treated, we could avoid the cotenability problem while deriving any and all counterfactuals allowed by our scientific theories. Now let us investigate a very simple example of counterfactuals which follow from a single set of facts and just one law. Take Newton's second law, $F = ma$. Instead of talking about 'hitting a', 'pushing a', etc., we now speak of a *net force* being exerted on a; and, instead of describing a's motion as falling or rising, we will note *a's acceleration* or its *velocity* (which includes its direction of motion). Describing what happened to a in these terms leads to the derivation of an infinite number of counterfactuals according to the above characterisation of laws.

Let's see how this works. First of all, we characterise body a as a Newtonian mass point, having M amounts of mass. From these few ontological preconditions, we can derive the truth of a set of propositions (infinite in number) about how a would behave with respect to acceleration were various net forces exerted on it:

1. $(X) \{(Nx.Mx) \supset (\exists f)[f: Fx{\to}Ax)]\}$, which reads 'For any x, if x is a Newtonian mass point and x has a mass M then there exists a unique mapping f between the net force exerted on x and x's acceleration.'
2. $Na . Ma$, i.e., a is a Newtonian mass point with a mass M. These are our only ontological preconditions. $\therefore$ ($\exists f$) ($f: Fa{\to}Aa$). Of course f is determined by the formula, $A = F/M$.

What does the above conclusion tell us about a's counterfactual behaviour? Given a has a mass of 1 unit, if a net force of 3 dynes were exerted on it, it would accelerate 3 cm/sec^2; 5 dynes would accelerate a 5 cm/sec^2; 10 dynes, 10 cm/sec^2, and so on. Using a few conditions and one of Newton's laws of mechanics, we can conclude all kinds of things would happen to a if a net force were to be exerted on it. Notice that the consequent of any counterfactual depends on the domain of f, *no matter what the actual net force on* a *may happen to be*. In fact, a's actual acceleration or any actual force exerted on it plays no role in the above inference schema whatsoever! – as it should be, according to the consequence theory.

What does the above simple case in Newtonian mechanics tell us about counterfactuals? In the first place, we now see that from a given physical system in a specific physical situation and the laws of physics we can derive an indefinite number of relations between a variety of physical variables, relations that inform us about the system's counterfactual behaviour. If we asked 'Which of these many counterfactuals actually holds for the system?' the appropriate response is to say '*They all do!*' The above schema easily accounts for this, for demanding that one of these counterfactuals has a privileged status would be like requiring only one solution to the equation, $Y = X^2$. Secondly, in spite of grammatical appearances, it is clear by now that these counterfactuals are not conditionals but are individual functional relations of the mapping f. This changes the entire character of Goodman's version of the consequence model and of cotenability. The traditional version has the antecedent of the counterfactual conditional conjoined with the conditions in order to derive the consequent. Because we are now no longer dealing with conditionals, i.e., what used to be called the 'antecedent' of the conditional becomes an *element*, Xn, in the domain and the 'consequent' is now the image of Xn, $f(Xn)$; now, Xn *does not partake in a truth functional derivation of* $f(Xn)$. Rather, the two are derived together. Because of this feature, exportation cannot be used to add the 'antecedent' to the conditions, thus removing one source of the cotenability problem. We will call our revised version the 'mapping model' of the consequence theory of counterfactuals, which has this simplest general form,

1. $(C_1' \ldots C_n') \supset [(A_1 \rightarrow C_1) . (A_2 \rightarrow C_2) \ldots (A_n \rightarrow C_n)]$
2. $C_1' \ldots C_n' \quad \therefore [(A_1 \rightarrow C_1) . (A_2 \rightarrow C_2) \ldots (A_n \rightarrow C_n)]$,

where each '$A_n \rightarrow C_n$' represents a functional relationship or a mapping between elements of the domain and range of f, i.e., $A_1 \ldots A_n$ are the elements of the domain of f, Xn, while $B_1 \ldots B_n$ are the elements of f's range, $f(Xn)$. (1 should be understood to be universally quantified while the conditions expressed in 2 are understood to hold for a particular system.)

As far as we are concerned, then, Goodman's original formulation of the problem of cotenability is, indeed, not well formed. If the reader will recall, it was pointed out that, in the match case, to use the truth of S and $\sim L$ to derive the rival counterfactual $S \supset \sim (O . D . M)$ was a highly suspicious move. On our scheme of things, this inference is clearly blocked. We can see this most readily once we realise that with our new version of the consequence model, Xn and *its image*, $f(Xn)$, *of the counterfactual are not part of the ontological preconditions of our law and, equally important, the ontological preconditions are not substituted in the invariable relationship expressed in the consequent*. So, for example, suppose we had body a suspended above the ground. Because it is a Newtonian mass, located in a gravitational field, albeit it is at rest and not accelerating, the law of freefall ($S = 1/2\, gt^2 + V_0 t$, which reads, 'The distance a body falls equals $1/2\times$ gravitational constant $\times$ the time of fall squared plus the initial release velocity $\times$ the time of fall) applied to a. Now the division of labour between the ontological preconditions and f should be clear. That a is at rest and is not being accelerated is part of the preconditions for the law of freefall. However, even though a is actually at rest and not accelerating, we can still infer, by the above schema, that 'If a were released with a velocity of, say, 10 cm/sec, it would travel about 5000 meters in 10 seconds.' (We also know we can derive an infinite number of other such counterfactuals, depending on t and V_0.) That a is actually at rest does not determine any counterfactual preference, and it certainly cannot lead to a derivation which falsifies any functional relation in f. In fact, suppose we fix $t = 0$, $V_0 = 0$ in the domain of f, which stand for a not being released and not accelerating. We then simply get $S = 0$ in the range, i.e., 'If a were not released, it would not fall', which is as should be expected. There are no inconsistencies here. Likewise, *no derived functional relationship can possibly contradict the set of ontological preconditions*. So, we no longer have the circularity problem of having to derive a proposition of the form $\sim [A_n \supset \sim (C_1' \ldots C_n')]$ in order to derive $(A_n \rightarrow B_n)$.

In contrast to this, the traditional version does not in the least distinguish between preconditions for the application of an invariant relationship to the system and the adverbial qualifiers which govern its behaviour. What the mapping model depicts as adverbial qualifiers are simply completing conditions for the traditional picture, conditions which can be added to $C'_1 \ldots C'_n$

in order to derive $C_1 \ldots C_n$. It is because of this that the fatal inference, $A_n \supset \sim (C'_1 \ldots C'_1)$ goes through.

Let us review why this inference is barred by the mapping model. Recall that Goodman was able to derive $A_n \supset \sim (C'_1 \ldots C'_n)$ from laws and conditions by means of a conditional proof which assumes A_n and then eventually deduces $\sim (C'_1 \ldots C'_n)$. This involved conjoining an A_n with a factually true $\sim C_n$, yielding $\sim (A_n \supset C_n)$ along the way. We then get $\sim (C'_1 \ldots C'_n)$ by *modus tollens*. This obviously will not work with the mapping model, however, for there are no propositions $(A_n \supset B_n)$ to negate in order to get $\sim (C'_1 \ldots C'_n)$. Now it may be protested that even if there are no $(A_n \supset B_n)$ counterfactuals in the mapping model, f nevertheless *entails* a set of conditionals, i.e., $[(A_1 \rightarrow C_1).(A_2 \rightarrow C_2) \ldots (A_n \sim C_n)] \Rightarrow [(A_1 \supset C_1) . (A_2 \supset C_2) \ldots (A_n \supset C_n)]$; and, when such an entailment is added to the mapping model, we get the cotenability problem all over again by means of hypothetical syllogism.

It is very important to see why the above will not work, and it goes back to the distinction between $C_1' \ldots C_n'$ and $f: Xn \rightarrow f(Xn)$ again. Since, according to the mapping model, we cannot substitute any $C_1' \ldots C_n'$ into the variables of f, any use of a precondition to refute an individual mapping relationship would be guilty of an equivocation. Following Sellars's distinction,[7] it is one thing to describe the state of a system before anything happens to it $(C_1' \ldots C_n')$ but quite another to depict *how* an action occurs to the system $(A_1 \ldots A_n)$. So, for example, one of the initial conditions allowing for the law of freefall's application to body a is that a is suspended in a gravitational field, i.e., a's actual velocity, V, equals zero. But that its velocity before release equals zero does not mean that we can plug in the value zero into V_0 of the invariant relationship depicted by the law of freefall, $S = 1/2 \, gt^2 + V_0 t$. That a's actual velocity is zero before release does not mean that its release velocity, V_0, is zero. In fact, even if the release velocity equals zero, *it simply is not the same velocity as* a's *velocity before it is released*, even though they equal the same amount, namely, zero. The value of V cannot be substituted in V_0 in $S = 1/2 gt^2 + V_0 t$ because it is not an adverbial qualifier which modifies *how* a is released. So we cannot combine the fact that a has not, in fact, moved with a is released in a conditional proof to derive $\sim$ (If a is released then it will fall). Hence, we cannot infer 'by means of a legitimate law' from the fact that a is not actually falling that releasing it implies that either it is not suspended in a gravitational field or that it is not a Newtonian mass. I will apply this very same reasoning to the match case in the next section.

The more generalised version of the problem is more troublesome because it does not appear to make use of this illegitimate 'plugging-in' of ontological preconditions into f. However, the more generalised version fails to distinguish between ontological preconditions, Xn and $f(Xn)$ as well. Recall the earlier difficulty confronting the consequentialist of adjusting to the addition of new conditions and handling how they would change the status of counterfactuals. Unlike the artificiality of Goodman's version of the problem, any version of the consequence theory must allow one to derive that the match will not be dry when struck if a moistening device is attached to it. And, we have seen that this can not be accomplished simply by adding this new condition to the original ones because it would also require the addition of a new law(s). We now know that this problem is easily solved once we make our laws adverbial, as captured by f. In other words, f does the job that the addition of a variety of new conditions was supposed to do.

In order to show this, let us take a legitimate case where it looks like the occurrence of the 'antecedent' of our conditional leads to derivation of the denial of the conditions. Suppose we were to discover that whenever a net force is exerted on a Newtonian mass point, its mass changes. This fact alone would cause cotenability problems for the traditional consequence model because the truth of the antecedent (the existence of net force) automatically implies the denial of the condition that a has a mass of M. Likewise, a counterfactual like 'If 3 dynes force were to be exerted on a, it would accelerate 3 cm/sec^2' would end up being false because the exertion of the force changes both a's mass and acceleration. This poses no problem with our new version of the consequence theory, however, thanks to the adverbial nature of our lawlike expressions. Once we know how forces change mass, we just pack this information into f instead of adding new conditions. So, for example, if the change of mass is inversely proportional to the square of the net force, we can adjust Newton's second law accordingly; a's new mass shows up in the range of the derived counterfactual but this does not in the least imply any contradiction between a's mass before and after the exertion of a net force. Thus, the solution to the general version of the cotenability problem is another case of keeping track of conditions, on the one hand, and the information contained in f, on the other. In other words, once we realise that the adverbial nature of laws renders them rich enough in informational content to enable us to derive an indefinite number of counterfactuals and that such a derivation takes place without tampering with our original ontological preconditions, the problem of cotenability is dissolved.

4. APPLICATIONS OF THE MAPPING MODEL

Returning to Goodman's match case, the chemical and physical complexities of the situation can overwhelm us. In fact, I submit that the match example is really misleadingly simple, for there is no state of the art when it comes to a complete, in-depth physical and chemical analysis of the behaviour of matches being struck under various circumstances. This is why I started out with the case of a single physical law and a few conditions telling us the counterfactual behaviour of a body a. The problem is not vagueness, as Lewis believes,[8] but that of applied science, specifically the application of scientific paradigms to everyday situations. At best, we can only describe what such an analysis might look like.

In order to adapt our new model to everyday situations, we must be sensitive to the nuances of scientific language that are often lost when carried over into everyday speech.[9] This is not meant as a condemnation of ordinary language – heaven forbid! – but a simple statement of fact that, for example, whereas we speak of tossing or releasing a ball, the physicist will speak of imparting such and such velocity, having net forces of such and such amounts exerted on a Newtonian mass, etc. Likewise, the scientist will replace 'striking a match' with 'a collection of sulphur molecules being moved along at a certain velocity, in contact with a surface having a specific coefficient of friction' and 'match lighting' with 'collection of sulphur molecules undergoing such and such amounts of rapid oxidation'. What good does this translation do? In the first place, *conditions become variables in equations (f) which carry the counterfactual information*. In other words, we are seeking an equation that will express the amount of oxidation of the match as a *function* of variables, among other things – and I am making all this up – the coefficient of friction of, say, the match box, the temperature of the match, the match's velocity and its normal force on the striking surface – all of this being in relation to the match's oxygenated environment, yielding an incredibly complex equation. In any case, f would look something like that.

Among some of the counterfactuals which would be derived from the ontological description of the system and our law(s) is that some ways the match is struck will be functionally related to its *not* being rapidly oxidised or lighting. For example, moving the match along the box at 1000 miles per hour may remove the surrounding air so that oxidation does not take place or our moistening device will cool the match as it is struck so that the degree of oxidation is well below that of rapid oxidation. Once more, this information is packed into our function, f. The point of all this is that we need some sort of 'dictionary' between the highly technical terms of f and ordinary language in order to apply our model to everyday cases. But it can be done, it seems, with very little difficulty. For example, if a's being released in a gravitational field happens to cause a table to be moved below it exactly when it is released, this is just another case of the net force on a being equal to zero, which means that if a were released in this way, it would not accelerate, i.e., it would not fall. But this is perfectly compatible with a being released *another way* and falling; and, *all* of these, of course, are compatible wih the ontological conditions of a being a Newtonian mass, suspended in a gravitational field.

So far, the move has been to maintain the adverbial content in the invariant, functional relationship of the law rather than in the conditions, the latter being done by the traditional consequential-ist. By doing so, some important logical features of counterfactu-als can be accounted for without having to appeal to possible worlds. Take, for example, this valid truth-functional inference: $P \supset Q \therefore (P . R) \supset Q$. However, from the truth of 'If I were to walk on the ice, it would not break', we cannot infer 'If I and my pet elephant were to walk on the ice, it would not break.' So how can the consequence theory be correct, for if $P \supset Q$ follows from laws and conditions $(P . R) \supset Q$ does as well? Let us see how the revised version of the consequence theory easily resolves this. Instead of expanding the initial conditions in order to accommodate each new situation, we pack the information into the invariant relationship again. Think of the ice as a molecular lattice, write down the Young's modulus (which is a measure of its rigidity) for the ice and arrive at some functional relationship between the forces exerted on this molecular lattice and its structural integrity. Crudely, we can divide the counterfactuals following from such an equation into two sets, one set dealing with forces equal to or greater than some number n while the other set involves forces less than n. If the force in question is equal or greater than n, the molecular structure fails, i.e., the ice breaks but not when the force is less than n. Again, the conditions of the ice being a molecular structure having a particular Young's modulus, etc., are the same throughout this variety of counterfactuals. So, my walking on the ice is translated into my body exerting a force on this lattice which is less than n, supporting 'If I were to walk on the ice, it would not break.' However, *the very same law and conditions also support* 'If I *and* my pet elephant were to walk on the ice it *would* break.' The truth of these two counterfactuals does not contradict our law and conditions. Although I cannot go into it here, the same strategy can be used to explain why and under what circumstances transitivity fails for counterfactuals. Another imaginary example brings us back to the general version of the cotenability problem. Suppose we wish counterfactually to describe the motion of a body, a, when released in a peculiar type of force field. Let us say that our laws tell us that a simply moves along a line (or geodesic) of the field. Before it is released the field lines are perfectly straight; however, let us add that releasing a in field causes the lines to bend. How, then, can we show that 'If a were released, it would move in a straight line' is false while 'If a were released, it would move along a curve' is true, even though the lines of the field are perfectly straight before a is released? According to the above strategy, the solution lies in describing *how* the lines of the field will warp as a function of the net forces exerted on bodies located in the field. Describing a being located within the field, without being released, allows us to apply our imaginary functional relationship between net forces and field warping to a, and we easily arrive at support for the latter counterfactual *vis-à-vis* the former, even though the lines of the field are actually straight.

5. CONCLUSION

We have seen how the distinction between ontological preconditions and adverbial constraints is the key to the solution of the cotenability problem which originally confronted the consequence view. What does this tell us about the nature of counterfactuals in relation to the workings of theory? In the first place, all of the possibilities of a system's future, unrealised behaviour are contained in the actual state of affairs, namely, the nature of the individual system, its relationship to other systems, etc. In other words, the possibilities of the system's behaviour, which is what counterfactuals are all about, are generated by its intrinsic features and its relations to the rest of the world. The chemical make-up of arsenic, for example, is such that according to our laws, we can project: if someone were to ingest such and such amounts of it, that person would become gravely ill or even die. Unless we take these ontological features into account, we will not have the faintest idea what laws apply to the system in question; we will not know which laws are even relevant. We have also seen in the previous sections how we should not confuse these ontological preconditions with the initial conditions of the D–N model; for doing so would lead, again, to cotenability problems.

Now we are in position to complete the counterfactual picture by showing how the support a theory confers on a counterfactual claim is a function of the common ontology it provides. Returning to the case where there is a glass of water before me, even if it were biologically possible for me to die were I to drink it, such a possibility would be an extremely remote one and such a fact would not lend support to 'If I were to drink it I would die.' In other words, if a theory supports a counterfactual claim 'If A were the case B would be the case', then the conditional probability of B given A must be high or equal to one.

This can be neatly captured by the mappings H and H^* of Chapter 7. What this means is that 'A' and 'B' in 'If A were the case B would be the case' refer to independent phenomena. Since counterfactual support involves inferring from the ontological preconditions that the probability of B given A is high or equal to one, such a claim is one of the mapping functions from the probability measures of our common ontology to the probability measures of independent phenomena. It is a member of H^*. But we know that H^* is induced or derived from H. A theory's counterfactual support, then, is tied in with its ability to reduce the number of independent phenomena. Thus, in 'If A were to occur then B would occur', at least two conditions must be fulfilled in order for our theory to confer support.

(1) 'A' and 'B' denote independent phenomena which are identified with aspects of a common ontology in such a way as
(2) to yield a high conditional probability for the occurrence of B, given A.

So, the major premiss of the above consequence analysis of counterfactuals which links the ontological preconditions with the adverbial constraints presupposes a common ontology which reduces the number of independent phenomena.

This completes the ontological turn in analysing theories. Not only is the common ontology provided by a theory responsible for its explanatory power, the way it fixes probabilities for combinations of events, but it is also behind the way theory provides counterfactual support.

NOTES AND REFERENCES

1. Nelson Goodman, *Fact, Fiction and Forecast* (New York: Bobbs-Merrill, 1955) pp. 13–15.
2. Ibid., pp. 16–17.
3. This version of the problem was suggested to me by R. Weingard.
4. D. Lewis, *Counterfactuals* (Cambridge: Harvard University Press, 1973) pp. 18–19.
5. For a strong defence and detailed examination of this approach to laws, especially in contrast to the truth functional approach, see F. Dretske, 'The Laws of Nature', *Philosophy of Science*, vol. 44, no. 2 (1977). While I agree with almost everything in Dretske's article, my use of mapping functions in the analysis of laws may or may not be in line with his thinking on the matter.
6. The distinction between ontological conditions and adverbial qualifiers can be traced to W. Sellars's 'Counterfactuals', in *Causation and Conditionals*, ed. E. Sosa (Oxford University Press, 1975) pp. 126–46. According to Sellars, once we distinguish between 'standing conditions', 'the *doing*' and the 'result of doing', we can not use the *fact* (= standing condition), say, that the match is not *now* struck or that it is not *now* lit to infer anything about what would happen to it were it struck (doing) under these conditions without pain of equivocation (pp. 134–7; 141–6). While I feel that his distinction is a sound one and that it is crucial to solving the cotenability problem, it is only a

partial solution, requiring its incorporation into the new model of the consequence view which I present below. Even if, as Sellars maintains, the laws connecting the antecedent and consequent of the conditional are restricted to doing and the result of doing, the conclusion is nevertheless a conditional. Its derivation, then, must have a conditional somewhere in the premisses, a conditional which allows us to deduce propositions of the form $A \supset \sim (C_1.C_2 \ldots C_n)$ again. Cotenability problems remain.

7. Ibid.
8. For those who are familiar with the literature on this, there is this difference between Lewis's analysis and mine. He believes that similarities among possible worlds play the essential role in determining the truth of a counterfactual while I believe counterfactual support rests on our theories about the *actual* world, no matter how commensensical or unsophisticated they may be. So, disagreement over the truth of a counterfactual may not reflect contrasting views of similarity among possible worlds, as Lewis believes (*Counterfactuals*, pp. 91–5), but that the disputants are simply working with rival theories about the world.
9. Ironically, although Lewis believes that counterfactual support is vague, he is in keeping with tradition when he allows for the assignment of a truth value to any and all counterfactual claims. On the other hand, I don't believe that this can be done willy-nilly. On the contrary, by my way of looking at counterfactuals, their truth is underdetermined by the facts and laws until the domain of *f* is precisely pinpointed. Otherwise, we should not know what to say about what would happen were certain things the case. Again, we must specify *how* things would happen.

10 Conclusion: An ontological approach to theories

The major objective of this text has been to formulate, develop and defend a theory about the nature of scientific theories. Part of what such a characterisation is meant to do is account for the variety of ways a scientific theory functions. Like nature, theories are many-faceted: they are responsible for a scientist's ability to explain and predict; they support the truth of certain counterfactual conditionals but rule out others; they are able to reduce or incorporate other theories, bringing about a greater unification within our conceptual scheme; and, finally, they confer meaning upon 'raw' data by providing us with a means of identifying and differentiating phenomena. The central thesis coming out of these chapters is that an account of the above aspects of theory can, for the most part, be traced to a theory's ontological commitments, its 'ontological zoo'.

Chapters 2–5 were devoted to providing the reader with a critical examination of classical approaches to the nature of scientific theory such as: logical positivism, Hempel's D–N model, Hesse's model analysis, Hanson's gestalt analysis, Kuhn's disciplinary matrices, Salmon's S–R model, and so on. While each of the above approaches has captured something very important about the nature of a scientific theory, I have also found them to be, at most, projections of a deeper, more structured view, that theories are first and foremost ontological depictions of nature. Why have all of the above analyses overlooked such an important and, in some ways, even obvious aspect of theory? The best answer, perhaps, is that talking about the ontological aspects of a theory as being most fundamental does not fit in very well with the traditional methods of analysis that were in fashion when the philosophy of science came of age in this century. Ever since the writings of Descartes and especially those of Kant, philosophy has gradually evolved in such a way as to place much – too much, I believe – emphasis on epistemological questions. Wolff refers to this development as the epistemological turn:

Such a picture [of the way philosophy was done before the epistemological turn] makes ontology prior to epistemology: *what is* is decided independently of *what we can know to be*.

After the epistemological turn, it becomes the reverse:

The realms of being and knowledge are coterminous, and even more significantly, the latter defines the former. What can I know? becomes the first and fundamental question of all philosophy.[1]

The epistemological turn has been carried to its logical conclusion by the positivists who insisted that ontology be completely subservient to epistemology: the only things that can be said to exist are those things whose existence is directly verifiable. The vestiges of the epistemological turn carried over, to some extent, in almost all of the above mentioned approaches to theory, where theories were thought of first and foremost as epistemological systems. After all, wasn't 'knowledge' supposed to be another word for science?

Another philosophical technique of analysis, which accompanied the epistemological turn, was the philosopher's concern to capture the logic of the language of scientific theory. Is it any wonder, then, when the task of analysing the epistemology of a theory is combined with ascertaining its logical and semantical anatomy,[2] that the above-mentioned philosophers have simply overlooked what I believe theories are actually all about. While I believe that capturing the epistemological, logical and semantical features of scientific theory is, indeed, a worthy intellectual pursuit, I feel that, if anything, misplaced emphasis on these features as rock-bottom determinants of a theory's essence actually gets in the way, that we must eventually come to an understanding of how it is that ontology shapes its epistemological, logical and semantical features. So this book can be considered an attempt to break away from these traditions and to turn the emphasis back to ontological considerations, though, of course, not necessarily in the way things were done in philosophy before the advent of the epistemological turn. If anything, I find the relationship between epistemology and ontology to be far more complex than many philosophers have previously envisaged, and I will not even attempt to give a complete description of it here.

Since each of the previously covered approaches to theory tends to accentuate a particular epistemological or linguistical feature, ranging from Hempel's logical syntax, Hanson's semantical levels to Kuhn's paradigms, we are left with the difficult choice of which school of thought comes to grips with the true nature of theory. Yet, if my thesis that the various epistemological, logical and linguistical features of theory can be traced to its ontological commitments, something I have argued for in Chapters 6–9, then the above controversy is dissolved. While each of the above-mentioned philosophers has hit on some important aspect of theory, it is nevertheless a mistake to identify that feature with the theory's essence. It has been my contention all along that they reflect a theory's characterisation of the nature of things and how it refers independent phenomena to an underlying common ontology.

Let us now cover how each approach accounts for the various aspects of theory in comparison to the ontological approach that was presented in the preceding chapters. Logical positivism traced a theory's explanatory power to the way it catalogued the facts, without any recourse to refer the facts beyond to theoretical entities. As logically elegant and purely empirical such an approach to science may be, it was found that the positivist's ontology was far too weak to account for the occurrence of laws going across categories. On the contrary, we have seen that predicting their existence is a rather *ad hoc* affair for the positivist, who restricts his ontological commitments to nothing but the facts and lawlike relationships among them. As a result, he cannot generate high probabilities for cross-category events while the realist predicts their occurrence in abundance. To the extent that such correlations are found in nature, the positivist loses out in prediction and confirmation to the realist. There is much irony here, for the above line of argumentation ends up 'turning tables' on logical positivism. What it says is that the positivist is not being empirical enough! Being wedded to an ontology which consists only of facts (independent phenomena) and laws which cover them actually ends up in limiting one's high probability predictions to correlations among independent phenomena alone, leaving nothing to be said about highly specific and sometimes fantastic relationships holding between completely diverse phenomena. Logical positivism, then, requires that our theories be empirical but if I am right about the above, the type of theory which meets their restrictions is not empirical enough to do the job. I will come back to this further on.

In a way, Hempel's D–N model is in a similar quandary, for even though he maintains that theoretical entities play an exceedingly important and useful role in explanation, their essential role is to supply one with premises in a deduction of the explanandum. But positivistic theories can deduce any set of phenomena which are derivational results of realist theories provided that the former can add enough empirical assumptions. It is just that they do not confer an equally high probability to cross-category events. So what reasons can Hempel give to prefer the latter type theory over the former except in terms of simplicity, i.e., theories which contain propositions about theoretical entities use fewer premises to deduce relationships among diverse, independent phenomena? It is true that Hempel tried to supply reasons for believing in theoretical entities in his analysis of the 'theoretician's dilemma'. The considerations there were mainly semantical, namely, showing that theoretical terms could not be reduced completely to observable terms. If the reader will recall, I felt that his argument based on the logic of reduction sentences was inadequate. There are better reasons for believing in theoretical entities, and they were supplied in Chapters 6 and 7. We know by now that these reasons presuppose an entirely different approach to explanation and the role theoretical entities play in explaining. Instead of being a means of deduction, they serve to explain by being objects of reference for independent phenomena.

Explaining by reducing the number of independent phenomena is amenable to a Hempelian analysis. Even so, I believe such an analysis is beside the point in that it fails to capture the essence of this type of explanation. What might such an analysis look like? I guess it would begin with H (our map from independent phenomena to aspects of a common ontology) and conjoin it with the laws covering the objects of the theory's ontology. Once Leibniz's indiscernibility of identicals is added to the argument, we can then deduce that certain phenomena are interrelated in terms of existence and the values of their variables. While I believe that this type of analysis can always be done, the real work of explaining is not the deduction but figuring out *how* independent phenomena are *referred* to the objects of a theory's common ontology in the first place. The real insight provided by the explanans is an ontological one. Recalling the Flatland case in Chapter 7, understanding resulted with the realisation that each circle was actually a manifestation of a sphere intersecting Flatland. It is true that once these independent phenomena are identified with different stages of the sphere's journey, one can deduce and predict a variety of relationships among the circles. But first, one has to solve the ontological jigsaw puzzle presented by nature's independent phenomena. This is not just a pragmatic residue of a deduction but the other way around. The deduction depends on H, an ontological claim about the way nature fits together. This is what explanation and understanding are all about. (The same holds, of course, even more dramatically for Salmon's S–R model.)

Not that all this should give contextualists such as Hanson and Kuhn reason to rejoice. While D–N has been rightly noted for its clarity and elegance, I am afraid that the same cannot be said for the contextualists. It is one thing to show that logical syntax or formal modes of analysis cannot get at the essence of a theory; and to this extent, I agree with the contextualists. But it is quite another thing to maintain that theories lack a formal structure worth analysing, that such contextualistic notions as 'theory-ladenness' and 'disciplinary matrices' do not require formalistic treatment. (It is true that Kuhn, as of late,[3] may have softened his opposition to formalising scientific paradigms.) Keeping in mind that the formal structure of a theory is indeed parasitic on ontological zoos in the manner described in Chapter 7, nevertheless, we know by now how difficult it is to use a gestalt model as a vehicle of expression for a notion such as theory-ladenness. Returning to the problems raised in section 4 of Chapter 4, I called into question whether Hanson could adequately explicate his ideas on theory-ladenness and, especially, semantic levels of language without utilising some kind of formal characterisation. The supreme irony, here, is that by placing so much emphasis on 'seeing as' and the application of a gestalt model to theories, like the logical positivist and the deductivist whom he criticises,

Hanson pays little more than benign neglect to the ontological aspects of a theory. In fact, at a key point in his chapter on causality, he balks at taking the final metaphysical plunge against his opponents by speaking of causality as, of all things, a linguistical-logical phenomenon, rather than an ontological one consisting of connections which actually take place in nature.

> Causes certainly are connected with effects; but this is because our theories connect them, not because the world is held together by cosmic glue. The world *may* be glued together by imponderables, but that is irrelevant for understanding causal explanation. The notions behind 'the cause x' and 'the effect y' are intelligible only against a pattern of theory, namely one which puts guarantees on inferences from x to y. Such guarantees distinguish truly causal sequences from mere coincidence.[4]

But this renders Hanson's notion of theory-ladenness a psychological-semantical one, leaving little or no connection between ontology and explanation. This does not seem to be any better than saying that theories are merely conceptual devices for the organisation of data, or placing guarantees on inferences, something the positivist would find to be entirely welcome. Let's face it. We like cosmic glue!

The account of 'theory-ladenness' and 'levels of language' implicit in $H–H^*$ presents an entirely different picture and, as expected, one completely alien to the gestalt approach. Recall that conceptual connectivity was the essence behind theory-ladenness. I do agree that the more theory-laden a system is the more conceptual connectivity it has relative to the others. But, as I stated in Chapter 4, I see no way to express conceptual connectivity within a science except to say that there exists at least one law or theorem of that science which connects the concepts in question. Otherwise, 'conceptual connectivity' functions more like an inexplicable primitive. It was also said that the source of conceptual connectivity is to be found in the ontology of a science. How does the ontology of a theory generate conceptual connectivity? The answer to this question should be apparent to the reader by now. Before independent phenomena are reduced by theory, we have seen how lawlike relations are not known to exist among the parameters of different independent systems. After reduction, however, it was noted how lawlike relations are induced across these systems, yielding a greater degree of conceptual connectivity than ever before. So it is no surprise, then, that conceptual connectivity is just another result of the common ontology provided by a theory.

This account of conceptual connectivity leads to a different version of theory-ladenness. Such a notion simply cannot be characterised in terms of the patterning of data. If anything, it is the other way around: concepts pattern data *because* they are theory-laden. But speaking in terms of 'patterning' is highly misleading in the first place. Admittedly, there is something in our story like a gestalt, where the whole is greater than the sum of the parts, but it is not a perceptual phenomenon. It is simply that the common ontology provided by theory contains more *information* than does the total of independent phenomena which it covers. For example, the concept of a sphere contains more information than a collection of circles. But this is because a sphere *is* more than a collection of circles *in virtue of its topographical structure* (its curvature). In other words, the common ontology provided by theory contains more information because the objects possess many more features, more properties and relations among them, and with a greater variety than do independent phenomena. This is what it means to say that the latter are mere aspects of a 'richer' ontology. So theory-laden concepts are those that refer to the properties and things of a theory's common ontology, which is represented in the range of H, while those concepts which are relatively less theory-laden are found in H's domain. Theory-laden redescription, then, amounts to shifting our description of phenomena from the domain to the range of H.

Once theory-ladenness is expressed in terms of a theory's ontology and H, Hanson's semantical notion of language levels at work in explanation falls out quite naturally. It was pointed out in Chapter 4 how one, very important, logical feature about levels of language was completely neglected by Hanson's analysis of degrees of theory-ladenness. Theory-ladenness entails some sort of a partial ordering among languages. If language$_1$ is more theory-laden than language$_2$, then it is not the case that language$_2$ is more theory-laden than language$_1$. Of course, a language cannot be more theory-laden than itself; and if language$_1$ is more theory-laden than language$_2$ and language$_2$ is more so than language$_3$, then language$_1$ is more theory-laden than language$_3$.

These asymmetrical, irreflexive and transitive features of theory-ladenness must be captured by any analysis worth considering. Hanson's gestalt approach never even attempts to do this. Yet these logical features of degrees of theory-ladenness are so easily captured by H. H is a homomorphism from the language of independent phenomena to that of a theory's common ontology. Imagine a series of languages which are mapped into one another. The more theory-laden a language is, the more other languages are related to this language as a domain of H.[5] The ultimate language with respect to theory-ladenness, if there is such a thing, would be one which serves as the absolute range for all other languages and is itself not mapped into any other language. The least theory-laden would be one where there exists no H having this language as its range. We must not forget how H is motivated by a theory's ontological zoo. In this way, semantics or levels of language recapitulate ontology.

Pretty much the same can be said for Kuhn's analysis of explanation being the assimilation of the explanandum phenomenon in terms of one's paradigm. Recall that a paradigm was a disciplinary matrix which contained four major elements. The first element, namely, symbolic generalisations and definitions, corresponds to D–N. But, according to Kuhn, much of the explanatory burden falls upon the second and third elements of the matrix, its metaphysical models and exemplars. Exemplars will be discussed later on but notice how these two elements are supposed to be the crux of his alternative to D–N. Kuhn supplies us with many interesting case studies in the history of science which are intended to show that it was the metaphysical beliefs – *vis-à-vis* laws as empirical generalisations – that played a key role in his understanding of nature, that it is in principle possible to find two formally identical theories which nevertheless yield different accounts of the same phenomenon because they are working under different metaphysical assumptions. The reader knows that there is much in this with which I agree. However, Kuhn never tells us what metaphysical models are, nor does he tell us how they work to explain in contrast to D–N. Until this is done, there is no clear alternative offered to the formalist. It is no wonder, then, that the most often found criticism of his notion of a paradigm is its lack of clarity. $H–H^*$ has attempted to make sense of the notion of a metaphysical model being primarily responsible for a theory's explanatory power, hopefully in such a way as to avoid any possible charge of vagueness from the formalist camp.

Understandably, the contextualists have very little to say about the reduction of one theory to a deeper, more fundamental system, except that it can not be done for reasons of meaning variation as we go from one theory to another. They contend the formalists falsely assume that the meaning of terms of the reduced and reducing theories are sufficiently univocal to allow a deduction of the reduced theory results from the reducing theory. Thus, they maintain, the formalists' deductions among different theories are always open to fallacies of equivocation. Now the formalist and the contextualist assume, in this dispute, that reduction takes place by means of deduction, a contention I have disputed throughout Chapter 8. The thesis in that chapter was that theory reduction is a variation of reducing independent phenomena, namely, when two predicates which belong to systems depicting entirely different phenomena end up referring to the same property or aspect of a common ontology. Thus, reduction takes place by means of an identity relationship which goes across independent phenomena. For example, kinetic theory has it that the temperature of an ideal gas and the mean kinetic energy of its molecules are one and the same property. That these two properties are related by identity is established by the fact that they behave exactly the same way with respect to all other properties. This feature has nothing whatsoever to do with the meanings of terms. On the contrary, it was insisted in Chapter 8 that if reduction is to take place correctly, the meanings of the terms in each scientific system are mutually independent. The version of reduction presented in that chapter characterises it as being *ontological* in nature: the number of distinct entities and properties in nature is pared. If reduction were, in fact, a form of deduction, we would indeed have to worry about equivocation in meaning as we go from premisses to conclusion. This is why meaning variation is an issue of contention between the contextualist and the formalist. But if reduction results in the way depicted in Chapter 8, by showing that diverse predicates of disjoint systems actually have properties in common, the above issue is dissolved; for the disputants end up being at cross-purposes because their mutual assumption that reduction takes place by deduction is false.

Although I feel that the formalist approach to theories has more to say about the nature of counterfactual conditionals than the contextualists, an analysis of these propositions cannot be a truth functional one because it is plagued with cotenability problems. In fact, it seems that the only way the consequence approach to counterfactuals can avoid these problems is to think of laws in some way other than truth functional terms. Contrary to what was believed by the logical positivists and advocates of syntactical approaches to theories, it appears that any account of how theories support the truth of counterfactual conditional claims will include some depiction of an ontology of nature. In the previous chapter, it was argued that the common ontology of a theory delineates the possibilities of behaviour, and that this is the source of counterfactual support. (In fact, I believe that counterfactuals in science are so interesting because they reveal such possibilities.) It was also contended that the support a theory confers on a counterfactual requires that the conditional probability of B given A in 'If A were to occur then B would occur' be high or equal to one. But this is a claim about the probabilities of the occurrence of A and B events as a function of the probabilities of the common ontology underlying them. Since such a claim depends on how a theory reduces the number of independent phenomena, counterfactual support is a function of a theory's common ontology. What this amounts to is that theories with different ontological commitments will support different counterfactuals, of course. I see no other way to account for how it is that theory establishes the truth of these claims, and such an account is conspicuously a non-truth functional one.

The above analysis of counterfactuals and how theory supports them leads to a new version of the structural identity hypothesis. Hempel's claim that the logics of explanation and prediction are the same has been severely criticised throughout this text. Likewise, Hanson's claim that there is an exact parallel between scientific observation and explanation has been faulted. If a structural identity hypothesis can be found, I believe it is this: a theory can explain B in terms of A if and only if it provides support for a counterfactual 'If A were the case B would be the case.' In other words, the very same aspects of a theory that are responsible for its explanatory power are also the source of counterfactual support, namely, its ontology. If I am right about this, then if a theory can explain why B occurs in terms of A's occurrence, it supports 'If A were to occur B would' as well – and vice versa. This is all well and good but the important thing to stress here is that this analysis of counterfactual support is not a truth functional analysis but a semantical one based on a set theoretical relation between the representations of a theory's ontology and independent phenomena. In the same way that the ontology of a theory determines the probabilities it assigns to events, it also determines counterfactual support.

We have learned that how a theory explains and predicts, supports counterfactual claims and incorporates other theories is a direct function of how it depicts nature. The last feature about theories which I will show to be parasitic on ontology is the classification of phenomena. Kuhn refers to this as the use of an exemplar in his disciplinary matrix. Let us review his analysis of theories as paradigms or disciplinary matrices. The reader will recall that Kuhn had the notion of an exemplar in mind in his original use of 'paradigm'. An exemplar served as a standard or a means of comparison for a variety of phenomena. In particular, exemplars enable the science student (or scientist) to extrapolate the laws of his science to a phenomenon which may appear to be entirely unlike anything experienced before. This is because exemplars have the power to inform the scientist that certain disparately appearing phenomena are actually of the same ilk. But how is the recognition that two differently appearing phenomena are the same accomplished? Kuhn apparently wavers on this. At first, he tries to use Wittgenstein's notion of 'family resemblance': the student takes notice of resemblances between the two phenomena in such a way as to warrant the application of laws in new circumstances. But this is precisely what exemplars are intended to explain. How is the student able to recognise their similarity? At first glance, a falling apple and the moon's orbit do not resemble each other in the least. Saying they resemble each other because of Newtonian theory begs the question. Later on, just like Hanson, keeping within the contextualist, antiformalist tradition, Kuhn falls upon a gestalt model of exemplars. The student simply 'perceives' that two phenomena are the same in the same way that different parts of a picture are patterned together. (In *The Essential Tension*,[6] he uses a gestalt-type example of an ambiguous figure being classified as a duck or a swan, depending on whether it is visually grouped with duck or swan background figures.) But how is this done? Well, according to Kuhn, it takes a lot of training! As far as I am concerned, this is not a good enough account; how the scientist identifies two differently appearing phenomena as the same phenomenon still remains a mystery.

An account of the identity of phenomena is forthcoming from my point of view but it requires that we drop the gestalt model. The reader should again be able to supply the story by now. Before a theory provides the student with a common ontology, independent phenomena are precisely that, namely, phenomena whose features and behaviour appear to have nothing whatsoever to do with one another. Let us use the Flatland case again as a point of illustration. Imagine a science student in Flatland who can predict the behaviour of circle events on the basis of a relationship depicting the radius of each circle as a function of time. Unfortunately, he is unable to predict the behaviour of ellipse events, as he has yet to figure out a relationship which covers them. In fact, he cannot perceive any kind of connection whatsoever between the occurrences of circles and ellipses; he is sadly unable to extrapolate from one to the other. Surely, circles and ellipses are not the same and it is natural to assume that as such they are subject to different geometrical laws. This difficulty is removed, however, when these two, diverse phenomena are provided with a common ontology: they are the *same type of phenomenon as conic sections* which result respectively from a sphere and an ellipsoid passing through Flatland. Now the very same laws of three-dimensional geometry apply to both the circle and ellipse events. We are now in position to transform from one phenomenon to the other and back again as a function of the curvature of the three-dimensional object which passes through.

The above is paralleled by another case from Chapter 7. Falling apples and the moon orbiting around the Earth appeared to be entirely different phenomena to pre-Newtonians. (Now how can one possibly 'gestalt' the two together, as Kuhn believes? What could possibly serve as a visual background which enables one to 'see them as' the same sort of phenomenon?) However, once apples and the moon are identified as collections of Newtonian atoms which are subject to universal gravitation, their terrestrial and extraterrestrial motions are different manifestations of the same type of phenomenon. As far as I am concerned, the only possible aspect of a theory which can function as an exemplar is its depiction of nature. Unless the science student is at home with the ontological commitments of a working theory, I suggest that he will have a most difficult time extrapolating from a familiar phenomenon to an entirely unfamiliar one. His explanatory and problem-solving repertoire will be severely limited.

The moral of my story should be clear to the reader by now. Change the ontological commitments of a theory, you then change the way it predicts and explains, the way it supports the truth of counterfactuals, how it classifies phenomena and the way it functions in several other ways. In this sense, the ontology of a theory is prior to the other aspects which have been stressed by the previously discussed philosophies of science.

While my defence of scientific realism rejects past approaches to the philosophy of science, it has a positive programme of its own. At the beginning of the book, I discussed how the logical positivist tried to get the philosopher of science to eschew any metaphysical concerns when it comes to presenting an analysis of the nature of scientific theory. We now know why this will not do, that the very workings of a scientific theory are based on the way it depicts the nature of things. If we are to arrive at an understanding of what theories are really all about, linguistical, logical and epistemological approaches, in themselves, will not take us there. On the contrary, if what I have been arguing for all along in this book is correct, philosophy of science must reorder its priorities. We have found that there is no way to practise science without being 'guilty' of doing metaphysics, without practising science from an ontological point of view. Showing this was a major objective of the argument for realism that was presented in the book.

If the above is correct, then new light is shed on the relationship between science and philosophy. Recall how one objective of logical positivism was to present a demarcation of legitimate philosophical claims from those of the empirical sciences: the former were solely about the logic and language of science while the latter were purely of a factual nature, with no room for metaphysical claims in between. But notice that the core of the realist's argument is to make the positivist's position an empirical one while using scientific progress as a means of testing it. If this can be done, and I see no reason why it can not, if it can be shown that the minimal ontology embraced by the logical positivist has empirical consequences, then the above dichotomy between philosophical claims and empirical ones breaks down. In other words, the argument for realism in this text commits a philosophical heresy: metaphysical claims which are about the nature of things have testable consequences, something the positivist tradition could not possibly accept.

Is metaphysics empirical? If the answer to this question is 'yes', then the line that divides philosophy and science is not so sharply drawn after all. Instead of a strict division of labour between scientific and philosophical activities, perhaps it would be better to speak of them diverging in some interests while sharing several others. Scientists and philosophers are nevertheless intellectuals who seek truth and understanding about the nature of things. At one time in history, there were students of nature who fitted the above characterisation such as Galileo, Descartes, Newton, Kant, *et al*. Many times, it was not entirely clear whether they were making considerations of a philosophical or scientific nature. They were called, aptly enough, natural philosophers. I think that it is an unfortunate outcome of logical positivism that this combination of activities fell into disrepute, bringing in the age of specialisation. But when Einstein asked about the nature of light or if the simultaneity of events was absolute or relative, were his questions of a philosophical or scientific nature? I think he was simply being a natural philosopher at the time, and that his questions, being ontological considerations, led to much scientific progress in the long run. Along with Earman, who perceptively pointed this out in his 'Who's Afraid of Absolute Space?',[7] I think that it would be better for philosophy and science to return to their interdisciplinary ways of the past, that an era of natural philosophy ought to replace the age of specialisation. In many ways, the arguments in this book add up to an expression of such recommendation.

If we philosophers are to get at the true meaning of a scientific theory we must work to glean its ontological implications, to seek out a view of nature that makes the most sense of its formal structure and predictive powers. As the student of philosophy will soon learn, especially in light of the rapid and mystifying developments which have occurred in recent physics, this will not be an easy task. It will probably take much more than a lifetime before philosophers comprehend the nature of things in light of the new quantum field theories. Until we gain ontological insight into these recent theories, unless we visit the subatomic zoo, their true meaning will escape us. This is fine, for it means that exciting developments in the philosophy of science lie ahead of us, I believe for generations to come.

NOTES AND REFERENCES

1. T. Wolf, *Kant's Theory of Mental Activity* (Cambridge: Harvard University Press, 1963) pp. 96–8.
2. 'The greatest weakness of positivism, in the philosophy of mind as elsewhere, is that it tries to make the notion of meaning bear too heavy a burden. This is always a bad tendency in analytic philosophy . . .' H. Putnam, *Philosophical Papers*, p. 451.
3. In Kuhn, *Essential Tension*.
4. Hanson, *Patterns of Discovery*, p. 64.
5. For example, ideally, the language of classical thermodynamics, which includes terms such as 'temperature', 'pressure' and 'entropy' can be mapped into the language of kinetic theory, with 'temperature' hitting 'mean kinetic energy of gas molecules', 'pressure' hitting 'average momentum transfer of molecules per area', 'entropy' going to 'degree of randomness', and so on. Everyday terms such as 'colour', 'solidity', 'elasticity', etc. would be embedded into the language of kinetic theory as well.
6. Kuhn, *Essential Tension*, pp. 308–19.
7. J. Earman, 'Who's Afraid of Absolute Space?', *Australasian Journal of Philosophy*, vol. 48 (1970).

Index

278   *Index*

DATE DUE

JAN 2 4 1985			